CW01668482

England's Vanishing WINDMILLS

A.E.P. Shillingford

Godfrey Cave Associates Ltd., London

End papers

The encampment at Blackheath in 1781 during the Gordon Riots in London showing one of the three mills which stood on the heath.

Frontispiece

Kennington Mill, Nr. Ashford in the 1930's. This mill has now disappeared completely.

First published in Great Britain by:
Godfrey Cave Associates Ltd.,
42 Bloomsbury Street,
London, WC1.

© 1979 Blaketon-Hall Ltd.

ISBN 0 906223 15 6

Colour photographs by:
A.E.P. Shillingford

Colour reproduction:
Kleur Litho, Bexleyheath, Kent.

Printed and bound in Great Britain by:
Hazell, Watson & Viney Ltd., Aylesbury, Bucks.

Contents

Acknowledgments

Much has been written on the subject of windmills, many books (now mostly out of print), articles in magazines, reports in newspapers, leaflets and booklets describing individual mills. The author of this present work readily acknowledges the debt that he owes to the authors of all this earlier material.

A considerable amount of information has also been obtained from talks and lectures given at the annual windmill meetings of the *Society for the Protection of Ancient Buildings.* In particular from the lectures of Mr. Rex Wailes who, by his writing and work, has probably done more than any other individual to promote present-day interest in windmills. It is not possible to mention everyone who has helped in the compilation of this material, but especial thanks are due to the following people and organisations:

Museum of English Rural Life, Reading
Abbotts Hall Museum of Rural Life, Stowmarket
Avoncroft Museum of Ancient Buildings, Bromsgrove
Manchester Art Gallery
Reading Public Library
Canterbury Public Library
The Newcomen Society
Ministry of Public Buildings and Works
The National Trust
Greater London Council
Bedfordshire County Council
Kent County Council
Drapers Mill Trust
Stansted Windmill Association
North Leverton Windmill Co. Ltd.
Eastbourne & District Preservation Trust Ltd.
Wiltshire Historic Buildings Trust Ltd.
The Herne Society Ltd.
The Burwell Windmill Trust
Uckfield & District Preservation Society
Pitstone Windmill Committee
Mr. Ablett, Framsden Mill
Mr. Jon Sass, Wrawby Mill
Mr. B.J.E. Briggs, Kibworth Harcourt Mill
Mr. R. Thomas, Outwood Mill
Mr. G. Thomas, Outwood Mill
Mr. F.W. Gregory, Nutley & Shipley Mills
Mr. R. Wootton, Herne Mill

England's Vanishing Windmills

Naked they leave the uplands green
Where once in bold relief they stood,
Lone and austere as might have been
The mills of God!

Shadows which vanish with the wind,
Songs with no echo in the hills...
Ghosts of the past, which leave behind
The legend of wind-driven mills!

Humble in aim, instinct with power,
Wondrous in craft, in form sublime,
Martyred by each inexorable hour
Of parted time!

By courtesy The Oxonian Press Ltd.

Introduction

Hidden in quiet corners of the English countryside or sometimes standing prominently amid the bustling modernity of less rural surroundings are many visual reminders of earlier and, perhaps, less hurried times. Market crosses, lock-ups and dovecots are good examples, but probably the most interesting of all is the windmill which, in common with its cousin the watermill, is a relic of an age when man harnessed the natural forces of wind and stream to grind his corn and drain his land.

From the crude windmills which first appeared in England around the 12th century, the English windmill gradually developed into a machine of some sophistication and along with the watermill made an important contribution to the rural economy.

In the early 19th century however a gradual decline began until, eventually, rendered obsolete by the advent of alternative forms of power, these attractive old buildings were, in most cases, abandoned to the destructive effects of time and weather. It has been estimated that as many as 10,000 windmills once stood in England, yet only a fraction of this number remain, and each year even this total is further depleted and the tottering shells of the many sail-less derelicts which remain portend the day when their dramatic silhouettes could be gone forever from the countryside. Few buildings can equal the appeal of the windmill, which, with its simple functionalism and sheer spectacular size, has for centuries made an outstanding visual contribution to the landscape. Fortunately many artists have recognised their beauty and many mills which have long since fallen have been recorded for posterity by pen and brush.

In recent years, thanks to the sterling

efforts of the *Society for the Protection of Ancient Buildings,* the nation has become aware of the threat to part of our heritage and many of the remaining windmills have been repaired and, in some cases, even restored to working order. This work must be encouraged and materially supported, for the windmill represents a phase of our technological progress, and an important period in the long history of corn milling, which began with the primitive quern and developed through the centuries to the present-day roller mill.

In this book an attempt has been made at an appraisal of some of the windmills which survive in England today with directions on how to find them. The mills which do remain are not by any means all mentioned in these pages, and it is hoped that the readers' curiosity will be sufficiently aroused to seek out others and investigate their history. An interest in windmills can be combined with other pursuits; motorists and ramblers for example can use a visit to a windmill as their reason to explore new territory.

Like artists, the photographer has found in the windmill a subject which demands recording. Those with an interest in the history of engineering will be attracted by the windmill with its ponderous yet ingeniously contrived machinery. The early history of the windmill is something of a mystery, its origin being somewhat obscure, but for the local historian, the history of the village mill usually provides a fascinating insight into the history of a locality.

Few windmills are working today, and those that are do so because of the enthusiasm of their owners, for the wind as a motive force is often unco-operative and therefore unsuited to the pre-set production targets which are so dear to the heart of the modern industrialist. With maintenance costs high, and craftsmen able to carry out repairs a disappearing race, it is not surprising that the day of the windmill as a viable economic proposition is over. The windmill today is a symbol not only of a former more leisurely age, but of man's early attempts at harnessing a natural source of power. In an era of rapid change and a pre-occupation with advancing technology the earlier inventions might easily be forgotten, but have we not a responsibility to future generations to ensure the preservation of some examples of one of man's earliest machines—the windmill.

The expertly restored smock windmill at Ibstone, Bucks. This mill was renovated for sequences in the film 'Chitty, Chitty, Bang Bang'. The sails have since been removed as the body was not sound enough to bear their weight.

Brief History

Although the date of the windmill's introduction into England has, with reasonable certainty, been established by documentary references, no reliable evidence has as yet been unearthed to prove from whence it came. Crude horizontal windmills were undoubtedly used in Persia during the 10th century, for they were mentioned by the 10th century geographer Al-Mas'udi in his work *Meadows of Gold.* These horizontal mills consisted simply of a vertical shaft carrying the sails in a square tower. Rotary movement of the shaft was obtained by directing the wind on to the sails via diagonally opposed slots in the four walls. There is no evidence however that this type of windmill was brought into Europe—in fact all pictorial representations of windmills on early manuscripts show the same type—the post mill.

An alternative theory that the Crusaders saw windmills being used by the Saracens and brought the idea back to Europe also has meagre evidence to support it, and it is now doubtful if the true origin of the windmill will ever be learned.

The Domesday survey of William the Conqueror was completed in 1086 and recorded over 5,000 mills, but it is unlikely that any were windmills. The majority were undoubtedly watermills, though probably many were worked by oxen or horses.

The earliest accepted reference to a windmill in England dates from 1185, and mentions the existence of one at Weedley in Yorkshire. Another well-known reference occurs in *Jocelyns Chronicle,* and refers to Dean Herbert's mill which he built on his Glebelands at Bury St. Edmunds in 1191, much to the Abbot Samson's displeasure.

As to which type these early mills conformed, we may only surmise, but they were possibly either fixed mills in which the sails were carried on an immovable body and therefore built permanently facing the prevailing wind, or perhaps sunk post mills, a type which had the trestle buried in the ground for stability. Neither of these two forms of mill have survived, but remains of the substructure of sunk post mills have been found here and there in England, notably at Sandon Mount, Herts. in 1934.

Three or four centuries after the first post mills appeared came the smock and tower mills, the earliest examples of both types carrying a single pair of stones. Whether the smock preceded the tower, or vice-versa, has not been established, but the earliest known remains of each type are the shell of a 15th century stone tower at Burton Dassett in Warwickshire and the smock mill at Lacey Green in Bucks., which was originally built at Chesham in 1650 and has recently been the object of a very praiseworthy restoration scheme.

The earliest known illustration of an English windmill is that depicted on a 13th century psalter. This tiny illustration shows what is undoubtedly an open trestle post mill with four common sails and a ladder. This manuscript, known as the *Windmill Psalter,* was bought by William Morris in 1896 and is now owned by the Pierpont Morgan Library in New York. Other illustrations include one in Aristotle's *Phisica* which was illustrated in England during the latter part of the 13th century, while yet another appears in the *Louterell Psalter* executed in the 14th century. Both of these priceless manuscripts are in the British Museum.

Centuries before the Domesday survey was carried out, grain was grown for food, and in very early times was prepared by pounding it between stones. Grinding the corn as an alternative to crushing it was first effected by the saddle quern, which consisted of a hollow shaped base stone, in which the corn was placed, across which another stone was drawn to and fro. The use of saddle querns was not confined to England; they were used world-wide and are even depicted in the decorations on the tombs of Egyptian Kings.

The rotary quern, which is still used in

many parts of the world, came some centuries later. It basically comprises of two round flat stones, the upper one being mounted on a peg fixed in the centre of the lower. By means of a handle the upper stone can be rotated to produce a continuous grinding action, the principle obviously being the one on which the revolving millstone (be it driven by animal, water or wind power) was based.

Grinding corn with querns was a daily chore carried out by each household and even when watermills and windmills had become firmly established the use of hand-mills did not entirely lose favour.

Manor Mills

During the middle ages every manor had its mill, either a watermill, a windmill, or sometimes both, and these were a profitable source of income for the lord of the manor. The ancient custom of milling soke, which was the right of the lord to compel his tenants to have their corn ground at the manor mill, ensured that he suffered no competition from other mills, indeed soke rights prohibited the erection of other mills in the manor. The obligation of the tenants to use the manor mill was not confined to the grinding of corn grown within the soke limits, but also to corn grown or bought outside. However, the manor mill was expected to be adequate for the needs of the tenants, with regard to both its grinding capacity and proper maintenance of the structure, otherwise the tenants were at liberty to take their grain elsewhere. Although the custom of milling soke was rigorously applied, it had never in fact been established by statute law, and the practice was, naturally enough, one of the bitterest subjects of dispute between tenants and landlord. Evasion of the soke 'laws' was commonplace, but the increasing reluctance of the courts to penalise offenders made it difficult for the lords of manors to enforce their 'rights', and the practice was eventually allowed to lapse.

The millers' payment or toll for grinding was a proportion of the grain ground, usually one-sixteenth. The measure used by the miller to appropriate his toll was known as the 'toll-dish' and carried the official manorial stamp. The grain acquired by tolls was the only grain the miller was allowed to sell and was stored in a special bin or 'ark'. Some millers yielded to the temptation to take more than the customary share and this led to the widely held view that all millers were rogues, a reputation which inspired an abundance of derogatory verse at their expense.

'Miller, miller dusty poll,
How many gristings have you stole?
One of wheat and two of rye,
Hang the miller six feet high'.

A further defect in the system of toll by grist was the fact that the proportion of grain taken by the miller was the same during a bad harvest as a good one, thus the payment for grinding was higher when the grain was scarce than when it was plentiful. This anomaly was eventually resolved when the payment of the toll in money only was established by law in 1796, a law which also effectively ended the widespread abuses of the earlier practice.

Windmills have on occasion been the scene of historical incidents. Richard, King of the Romans, was obliged to seek refuge in a windmill after the battle of Lewes in 1264. During the Civil War windmills played their part. In 1642 Charles I is said to have watched his first battle from a post mill on the top of Edgehill. That particular mill has been demolished in comparatively recent years, a loss which is especially regrettable in view of its historical connections. Curiously enough Charles I also witnessed his final defeat at Naseby from a windmill in 1645.

Nearer the present time, during both world wars, windmills have been called into service as observation posts, a point noted more than once elsewhere in this book.

A Large Beech Tree, William Todd, 1766.

Windmills on Maps

The windmill, being a distinctive landmark of some size, has for centuries been used by cartographers to embellish their maps, often with some artistry. Unfortunately many early maps are not as reliable as one might expect and although often claiming to be 'from the latest surveys by the best authorities' were in many instances copies of earlier maps with little, if any, revision.

'as to the several sets of county maps professing to be drawn from the latest observations, they are almost invariably copies of those that preceded them'.
(Richard Gough — *British Cartography* 1780)

John Speed's attractive map of Surrey dated 1610, for example, was re-issued again and again up until the mid-18th century with little revision, and yet had in fact been originally copied from John Norden's map of 1594.

The presence of a mill symbol on an early map does not necessarily establish the date of a mill still standing on that site. Disregarding the less than perfect accuracy of the scale on many old maps, which makes it almost impossible to pin-point an exact spot, many mills still standing today are the last of a succession of mills on that site, therefore a mill depicted on an old map could well be one which was later replaced. Indeed, in some cases, the mill shown had never existed—the mill symbol merely served, like the elephants on innumerable maps of the African Continent, to fill up an otherwise empty space.

However, maps based on fresh surveys were drawn up from time to time and, as long as the information shown on them is regarded with caution, their study can be a useful and fascinating preliminary to research into the history of any particular mill or mill site. Even on the present-day excellent O.S.I. maps anomalies exist, and seekers of windmills must sometimes suffer the disappointment of finding a 'desirable residence' where, according to the map, there should stand a mill.

Dated Inscriptions

Many windmills have a date carved somewhere on their woodwork and these are often accompanied by initials. These early inscriptions, although of great interest, are not very reliable as a means of dating the erection of a mill. Many mills were built on the site of earlier mills which were in need of replacement. In such cases, sound

Holton Mill, Holton St. Peter, Suffolk — Built in the mid-18th century, this attractive post mill is one of several in the county which have been rebuilt in recent years.

Brill Mill, Bucks. — This post mill, the sole survivor of three similar mills which once worked here, was built in the 17th century. The photograph was taken during rebuilding work and the mill has since had the framework of the sails fitted.

Great Chishill Mill, Cambridgeshire — This fine post mill, rebuilt in 1819 from an old post mill dated 1726, is one of the few post mills which survive with an open trestle. Others stand at Bourn, also in Cambridgeshire, Chillenden in Kent and Nutley in Sussex.

Tower Mill at Cley-next-the-Sea, Norfolk — Dated 1713, this mill ceased grinding about 1917 and was converted into a house shortly afterwards.

Stracey Arms Mill, Norfolk — Dated 1883 this fine tower mill stands beside the river Bure.

Drainage mill at Horsey Mere, Norfolk — Built in 1912 this nicely preserved mill is owned by the National Trust.

A common sight in Norfolk are the now disused drainage mills, most of which are derelict. This one stands beside the river Bure.

The Fen district round the Wash was for centuries an inland sea in winter and a dangerous swamp in summer. In 1629 James I summoned the Dutch engineer Cornelius Vermuyden to drain the land. This cross-section of a drainage mill on the Dutch plan shows how the breast wheels raised the water, by means of gearing, and discharged it into the river.

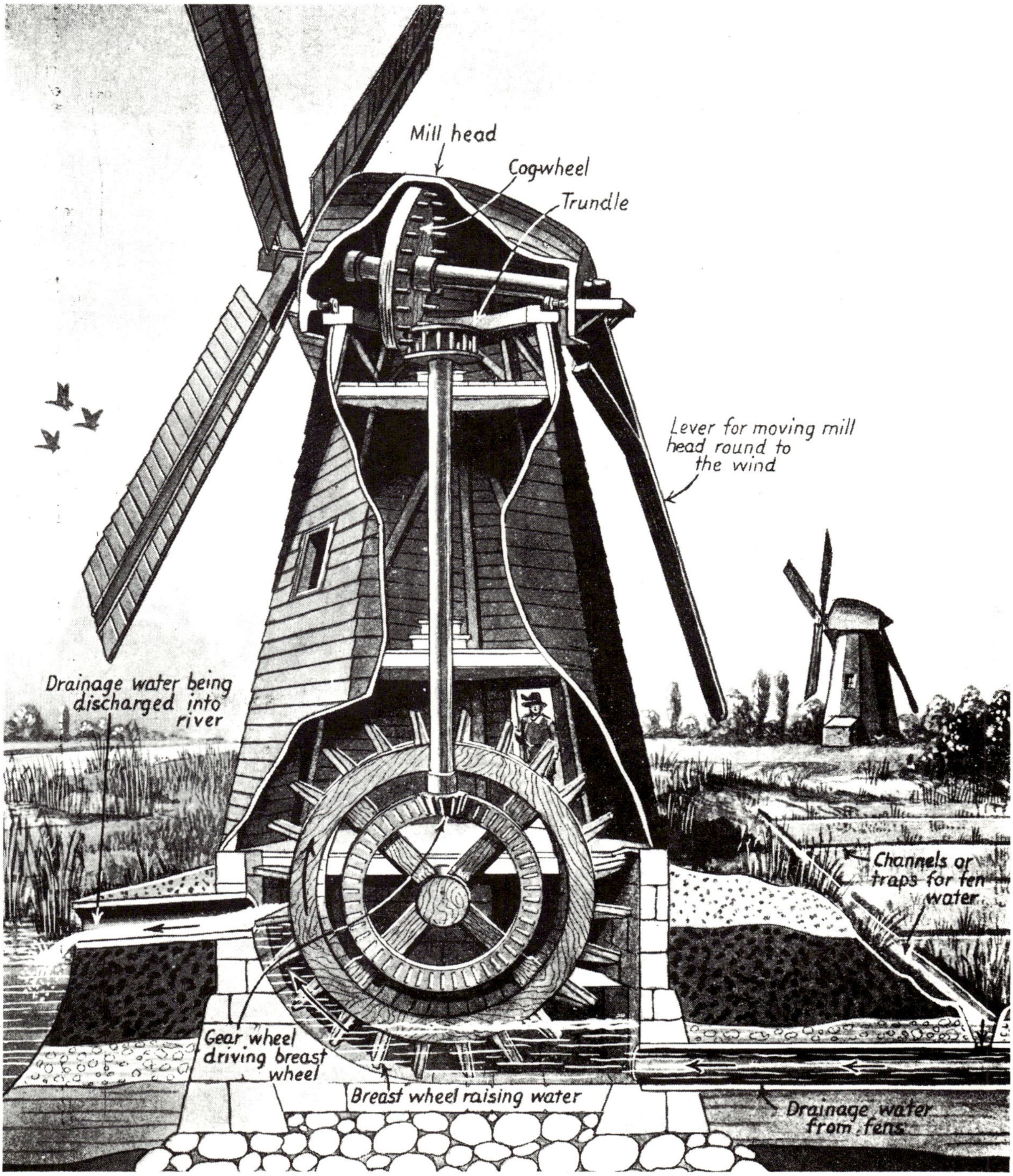

The Decline of the Windmill and Sources of Alternative Power

The application of steam power to milling took place in the late 18th century. The first steam-driven mill was built in 1781 at the Victualling House of the Navy at Deptford, and its example was followed shortly afterwards by the Albion Mill at Blackfriars, which was erected in 1784. Although acceptance of these large and efficient mills by the millers was slow at first their advantages were quickly proven. They were able to work regardless of wind and weather and could be sited anywhere, in fact many were built at ports to receive imported grain straight off the ships, and there was virtually no limit to the number of stones that could be driven.

The decline of the windmill had begun, a decline which was accelerated by advance in land transport. The rapid growth of the railways and the arrival of motorised vehicles enabled grain and flour to be carried to and from these large mills to all areas and, as a result, many of the country mills which had previously provided an essential service to small local communities had to close, since they could not provide flour at competitive rates. Many mills were fitted with small auxiliary steam engines to power the stones on windless days, and in this way many a hitherto unprofitable mill was given a new lease of life. Sometimes when a mill required expensive repairs to its sails or wind-driven machinery, its auxiliary engine would be brought into use permanently to grind the corn.

Other industries also made use of the steam engine, and the consequent increase in the demand for labour to man the factories enticed large numbers of countryfolk from the rural areas. Men who were able and willing to work the irregular hours necessary to wind-milling became increasingly difficult to find.

The steam engine was later followed by the development of the internal combustion engine powered by gas or oil and, eventually, most convenient of all, the application of electric power.

Roller milling, first successfully carried out in Switzerland in around 1840, was introduced into England in the 1870's and provided additional serious competition to the windmill.

The First World War (1914–18) saw the cessation of flour production by a large number of windmills. The enforcement of government regulations controlling the production of flour was one factor which contributed to this recession. A further significant reason was that new bolting cloth for the flour grading machines was unobtainable. The difficulty of maintaining existing bolting cloth in good condition due to attack by moth, added to normal wear and tear, made the task of keeping a windmill operating economically virtually impossible.

One of the factors contributing to the decline of corn-mills, steam power, was also deposing the marsh mill. Attempts at reclaiming the Fens had been carried out at one time or another since Roman times. Although it is known that wind powered drainage mills were at work in the 16th century, it was not until James I engaged the Dutch engineer Cornelius Vermuyden, who arrived in England in 1621, that any degree of success was achieved. Many hundreds of windmills were used in the reclamation of the Fens and marshlands, but with the introduction of at first steam, then oil engines, and later electrically powered pumps, the windmills were abandoned one by one.

Structure and Machinery

The purpose of a windmill is, fundamentally, to convert the horizontal movement of the wind into circular motion to drive machinery. That this circular motion is initially obtained by means of sails is immediately obvious to the most casual observer. To obtain maximum driving power however, it is essential that the sails are kept facing squarely into the wind, no matter from which quarter it may be blowing. The manner in which this is done separates windmills into two distinct types—the earliest, *the post mill,* is based on a principle in which the entire mill is rotated. Alternatively the sails can be carried on a smaller, and therefore lighter, part of the structure, which can be rotated independently of the main body containing the machinery. This makes the operation of 'winding the mill' less cumbersome and led to the introduction of the second and more advanced type of windmill, *the tower mill.*

A rare survival — the compass arm wallower in Lacey Green smock mill sillouetted against the sky after the cap and windshaft had been lifted off.

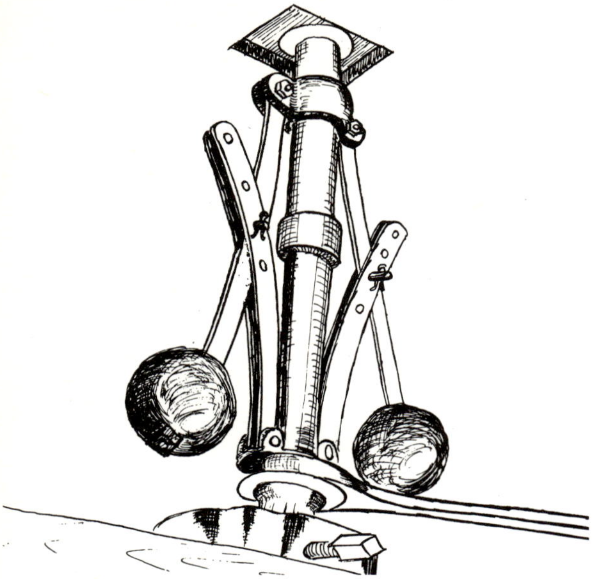

Governor: Outwood Mill

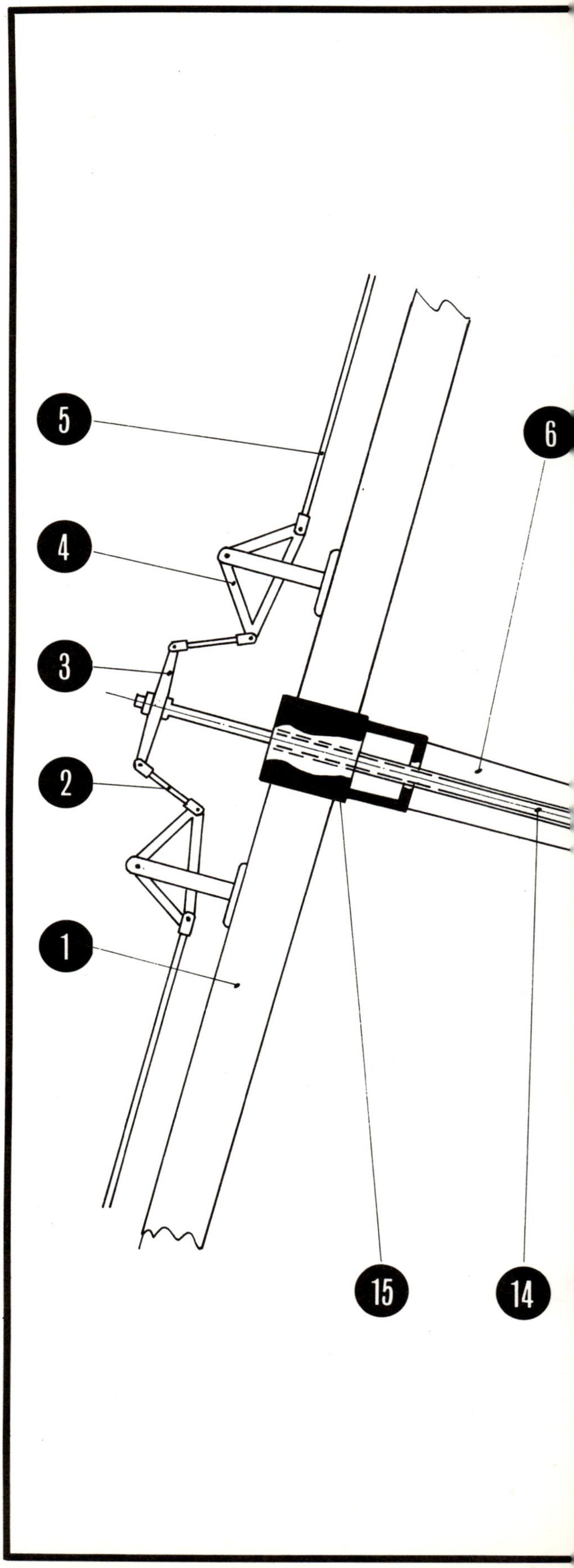

Diagram of Patent Sail Striking Gear:
In operation if the wind opens the shutters the movement of the shutter bars (5) will, via the bell cranks (4) connecting links (2) and spider (3) actuate the striker rod (14) which passes through the hollow windshaft (6). The coupling (9) connects the striker rod, which rotates with the windshaft, to the rack (12) which does not rotate. Movement of the rack will rotate the pinion (11) which is mounted on the same spindle as the 'Y' wheel (10). The chain (13) running over the 'Y' wheel is hung with weights, thus the shutters will open only if the wind strength is sufficient to lift the weights selected by the miller to suit the prevailing wind conditions.

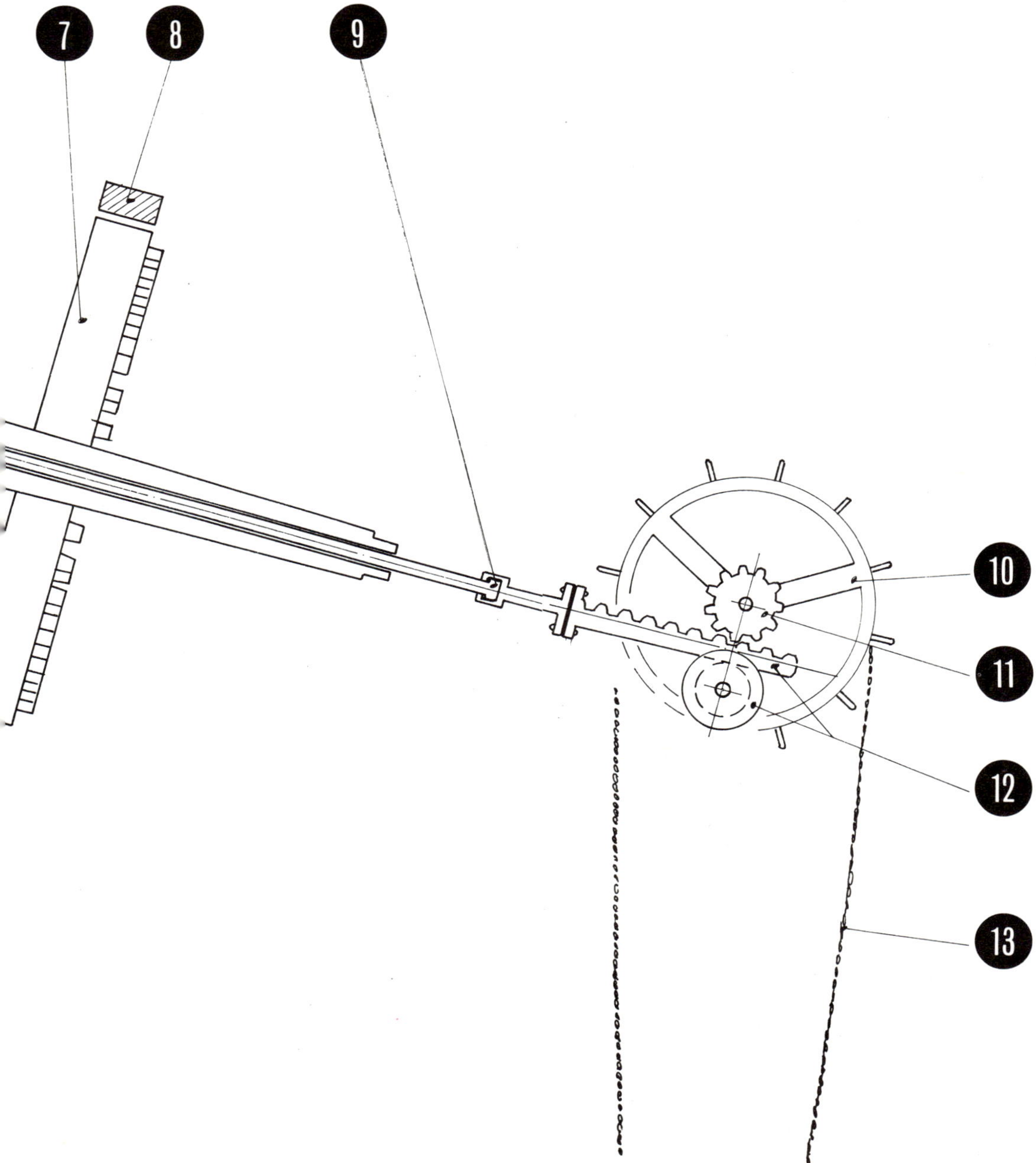

7
8
9
10
11
12
13

The Sails

The dominant feature of a windmill is, of course, its sails—even from afar their distinctive outflung attitude gave to the windmill an outline that was unmistakable—unique. With the exception perhaps of the annular sail or windwheel, introduced when windmilling was on the wane, the basic shape of sails changed very little over the centuries. Technically, however, sail development progressed from the simple cloth covered framework of the medieval mill to the latter day patent types which are an impressive example of the remarkable ingenuity shown by millwrights.

The sails of the earliest mills consisted of two stocks of timber mortised crosswise through the projecting end of the windshaft. Simple wooden frameworks were attached to each stock and sailcloth was threaded lengthwise in and out of the sail bars. The weather of these sails was constant; that is the angle at which they were set to the wind was the same throughout the length of the sail. The speed of the tip of a sail is however much faster than the inner end and the early millwrights found by trial and error that a twist or 'weather' over the length of the sail dramatically improved its efficiency. Individual millwrights applied their own 'know how' but generally the weather on a typical sail would diminish progressively from about 25 degrees at the inner end to almost zero degrees at the tip.

The medieval sails were gradually improved over the years and eventually evolved into the common sail still to be seen on some restored mills (Pitstone Green is an excellent example). The common sail retains the sailcloth, but this is spread across the face of the sail. Sails became larger by lengthening the stocks and the addition of whips, smaller section timbers clamped to the stocks. The sail bars were mortised into the whips, the angles of the mortices varying over the length of the whip to achieve the correct weather.

The area of the cloth on the common sail had to be chosen to suit the wind strength, and in variable wind conditions much time would be wasted stopping the mill to increase or decrease the area of canvas. Each sail had to be brought into the 'six o'clock' position to make the adjustment and the miller might have to climb the sail, perhaps in pouring rain. Hair raising incidents, caused by sudden failure of the brake to hold, were by no means uncommon, many a miller having involuntarily gone round desperately clinging to the sails.

The spring sail, which was invented in 1772 by the Scottish engineer Andrew Meikle, disposed of some of these disadvantages. A mill still had to be stopped with each sail positioned in turn near the ground but adjustment was much easier and quicker. With this type of sail the sailcloth is replaced by a row of hinged shutters set along each sail at right angles to the whip, the shutters were either made of wood or light wooden or metal frames covered in canvas. All the shutters are interconnected by means of a long bar running the length of the sail which is spring loaded at the windshaft end. The spring tension is set by a screw or hook and notch device located at the sail tip, adjustment being made as with the earlier sail at the 'six o'clock' position. When running, if the wind strength increases sufficiently to overcome the spring tension set by the miller, the shutters open and 'spill the wind'. Spring sails are not however as powerful as the earlier cloth sails, and it became common practice to equip a mill with a pair of each type and so derive some of the advantages of both.

In 1807 the patent or self-reefing sail designed by William Cubitt appeared. This was a direct development of the spring shuttered sail and was born of the need for a sail which could be adjusted without stopping the mill. The springs are dispensed with—instead, each shutter bar has a bell crank lever fitted to its inner end and these are coupled to the arms of a cruciform iron spider mounted on the fore-end of the iron striking rod which passes through the hollow windshaft. The tail-end of the striking rod projects from the rear end of the windshaft and is fitted with a toothed rack which meshes with a cogged wheel carried on a spindle Mounted on the same spindle as the cog-wheel is a pulley carrying an endless chain. A pull on the chain will move the striking rod back and forth causing the shutters to open and close. Weights are hung from the chain to hold the shutters in the closed position and when the mill is working the shutters will open if the wind strength increases sufficiently to raise the weights. Adjustment to the amount of weight holding the shutters closed can be made while the mill is running.

The majority of mills carried four sails, although five, six and even eight sailed mills were built. Many mills were run with just two sails, for if either of the other pair were damaged, both the defective sail and its opposite number would have to be removed to maintain the balance essential to smooth working. Thus a mill with an even number of sails had an advantage over a five sailer which, if damaged, was effectively put out of action until the damage was repaired.

The annular sail or windwheel, now familiar on modern wind pumps arrived too late to make much impact in England. A well-known example was on the tower mill which stood at Haverhill in Suffolk.

Windshafts

The early windshafts were usually made of heavy section oak, a diameter of about two feet being quite normal. The sail stocks were at first carried through mortices in the shaft, a method of construction which was unsatisfactory due to water penetration which caused the shaft to rot. Immediately behind the sails was the neck bearing, made of hard wood on the earliest mills, but on later mills other materials were used including brass, stone and marble. Wear on the wooden shaft, where it rested in the neck bearing, was minimised by recessing strips of wrought-iron into the timber. An iron pintle was inserted into the tail-end of the windshaft, and this located in a simple thrust bearing. On the later mills windshafts were made of cast-iron and were much smaller in diameter, with the sail box or poll end cast integrally. Many early wooden shafts, perhaps with defective mortices, were modified by having the front end sawn off and a cast-iron poll end fitted. The poll end consisted simply of two hollow square boxes cast at right angles into which the stocks of the sails fitted, held in place with wooden wedges. An alternative to the poll end, commonly used in the north-east, especially on multi-sailed mills, was the iron cross, a massive cruciform casting to which the whips of the sails were bolted or clamped, sail stocks in this case being dispensed with.

At the forward end of the windshaft immediately behind the neck bearing was the brake wheel, a large cogged wheel common to all windmills and the first in the train of gears which transmits the rotation of the sails and windshaft assembly to the machinery below. As with windshafts the early brake wheels were commonly made of oak with each cog separately mortised in, apple wood being accepted as the best material for cogs. On the earliest brake wheels, compass arm construction was used in which the wooden spokes or arms were mortised into the windshaft. This form of construction was later supplanted by the much stronger clasp arm arrangement which has the arms in parallel pairs set crosswise to form a square aperture in the centre. A square section is provided on the windshaft on which the brake wheel is mounted, held and centred.

In the late 18th century with the increasing use of cast iron, the majority of the new mills either had cast iron brake wheels or composite types of timber and iron. Cast-iron did not entirely supersede timber for the construction of cogs in windmill gearing because iron cogs meshing with wooden ones gave a much smoother and quieter running drive. The brake itself was carried around part of the circumference of the brake wheel (hence its name) and was made up of a number of separate segments, usually elm, which were hinged together in the form of a band. One end of the band is pinned to a fixed timber and the other is attached to a heavy lever. The weight of the lever, if allowed to hang free, will contract the band around the rim of the brake wheel. To start the mill (assuming the wind is blowing) the brake is released by raising the lever by means of a rope passing over a pulley, the rope hanging down to some convenient point where the sails may be seen. By raising the lever slowly, a pin in the side engaged with a swinging hook which holds the lever up in the brake off position. A sharp pull on the rope would disengage the hook to re-apply the brake. Great care was required in applying the brake to stop a fast running mill as the friction generated could, and sometimes did, set fire to the mill.

Winding the Mill

Until the middle of the 18th century windmills were luffed into wind manually. The original method was by means of a tailpole and on post mills this often had yokes fitted on the end against which the miller would place his shoulders to push the mill round. In some cases the end of the tailpole was supported by a cart wheel and the mill was winded by hitching a horse to it and pulling it round. Alternatively a number of stakes were driven into the ground in a circle around the mill and a small hand winch fitted to the end of the tailpole. A rope or chain would be played out and tied to a suitable stake and the mill winched round a bit at a time, the end of the rope being moved from one stake to the next. The early tower mills were also winded by tailpole but it was less satisfactory, the steep angle of the tailpole from the cap to the ground provided little leverage and made the task of rotating the cap far from easy. The introduction of chain operated winding gear on tower mills made the task much less cumbersome, the miller being able to rotate the cap by simply pulling on the chain, the operation being carried out either from the ground or a stage built around the mill. A means to bring the sails into wind automatically, the fantail, was introduced in 1745 by Edmund Lee. It consisted of a small wooden bladed fan set at right angles to the sails at the rear of the cap. When the sails are facing square to the wind, the fantail is edge on and therefore stationary, but if the wind veers and strikes the fan at an angle, it sets it in motion, which through a system of reduction gearing brings the cap slowly round until the sails are facing into the wind. Fantails and their attendant gearing were so finely adjusted that any slight change in wind direction set the gear in motion, always keeping the sails facing the right quarter.

The Stones

The actual process of grinding is carried out between pairs of stones, each pair consisting of a lower fixed bedstone, set perfectly level by means of wedges in a frame in the stone floor, and a runner stone which rotates above it. Encircling the stones is a wooden casing or vat to contain the meal as it issues from between the stones. A typical millstone would measure about four feet in diameter by a foot thick and weigh about three-quarters of a ton. Despite its weight the runner stone had to be finely balanced and free running, otherwise when spinning at the usual grinding speed of around 125 r.p.m. the vibration caused by poor balancing could cause rapid deterioration of the stone spindle bearing and possibly structural damage to the mill. Balancing a runner stone was carried out by pouring lead into pockets arranged around the outer edge; on some later mills runner stones were fitted with patent adjustable balance weights. The runner stone is supported on a spindle which passes down through a bearing set into the centre of the bedstone, the lower end of the spindle locating in a bridging box on the bridgetree. The bridgetree is a stout pivoting beam which provides a means of raising or lowering the runner stone in relation to the bedstone and thus adjusting the size of the gap between the working faces. This operation is known as tentering and is described later. The bridging box enables the bearing carrying the lower end of the spindle to be moved sideways in any direction, by means of adjusting screws, and ensures that the spindle can be set exactly vertical, which is essential if the runner stone is to run truly level in operation. The top of the stone spindle, where it protrudes through the bedstone, is shaped in the form of a tapered square. Located on this tapered end is a small iron casting known as a mace. Let into the lower end of the eye of the runner stone, and held in place with lead or wedges, is an iron gimbal bar which locates in a groove in the top of the mace, the fit being slack enough to allow a well balanced stone when running to be self levelling.

Several types of stone were used for milling, but in most of the windmills which survive today with the machinery and stones intact, both peak stones and French burr will usually be found. The former is a hard sandstone quarried in Derbyshire and used for grinding the coarser grain, such as oats and barley. French burr was used for wheat, and whereas the other types of stone were quarried in one piece, French burr stones were made up of small pieces of stone shaped and matched, set in cement, and then backed with plaster of paris, the whole assembly finally being bound with iron hoops to hold it together. Other types of stone used were the imported cullin stones originating in Cologne and the later emery composition stones—these were less favoured however. In addition to Derbyshire, stone for grinding was quarried in Sussex, Buckinghamshire, Herefordshire, Monmouthshire and Lancashire—doubtless there were other sources too.

The working face of each stone has a pattern of furrows cut into it running obliquely from the eye. This pattern is identical in both runner and bedstones if the stones are compared side by side. When the runner stone is set in place over its counterpart, the patterns are reversed and the action of a furrow in the revolving runner stone passing over a furrow in the bedstone is 'scissors-like' and cuts the grain. This effect is repeated by the system of fine cracking in the lands between the furrows and combined with the secondary purpose of the furrows, namely to act as channels, serves to work the meal towards the outer edge of the stones whence it falls into the vat.

Lacey Green Mill, Bucks. Iron pintle let into top of the upright shaft. In the background can be seen the iron curb and rollers on which cap turns.

Above: Lacey Green Mill, Bucks. The windshaft and compass arm wheel removed from the mill for repairs. The brake can be seen around the rim of the brake wheel. The neck bearing for the windshaft can be seen in the left background.

Stone Dressing

To ensure efficiency of grinding, it was necessary to redress the stones at frequent intervals, because the cutting edges of the furrows were quite quickly dulled. In the heyday of the windmill itinerant stone dressers travelled from mill to mill dressing the stones in each on a regular basis. The work was highly skilled and before a miller would entrust his valuable stones to an unknown stone dresser, it became customary to ask the man to 'show his steel'—small splinters from the mill bill which became embedded in the skin on an experienced stone dressers hands.

The eventual decline of the windmill resulted in a scarcity of such craftsmen and experienced stone dresser's hands.
himself in the handful of working mills which remained. In addition to skill, stone dressing involved very hard physical labour—a runner stone weighing perhaps half a ton or more had to be lifted off the stone spindle and turned over. This was done by driving in wooden wedges and levering the stones apart until a gap was obtained large enough to pass a rope between them and through the eye. The stone was then hoisted with either block and tackle or a rope around the windshaft and turned over onto a sack of corn or something similar. After dressing, the stone was hoisted back onto its spindle and rebalanced, it being customary also to grind a bag or two of grist, both to check the work just carried out and to remove any loose grit lodged in the stones.

An interesting point regarding the wearing down of the stones during usage is the fact that the particles worn off added a small and apparently beneficial amount of calcium to the flour. This lack of calcium in the product from the later steel roller mills was thought to be a cause of rickets, and regulations were enforced to provide for a proportion of calcium to be added to all flour from roller mills.

Facing page:
Lacey Green Mill, Bucks. Close-up of iron poll end showing iron bands which clamp wooden windshaft around the cross tailed gudgeon.

The Drive to the Stones

The drive to the single pair of stones in the early mills was direct from the brake wheel on the windshaft and this form of drive still survives in many old post mills. The drive from the brake wheel is via an iron spindle known as a quant, the upper end of which is carried in a bearing or glut box mounted on a horizontal sprattle beam. The lower end of the quant is forked and locates in slots in the mace, a small casting fitted on the stone spindle. Keyed to the quant just below the glut box is the stonenut, a cogged wheel made of iron or wood, which meshes with the brake wheel. The location of the quant in the eye of the stone has sufficient clearance to allow the upper end carrying the stone nut to be moved sideways out of the glut box to disengage the drive. Direct drive, although simple, had the serious disadvantage that it restricted the layout of the stones and thus limited the number that could be carried. Indirect drive disposed of this drawback. With this type of drive the motion is transmitted from the brake wheel by a large bevelled cogged wheel known as a wallower. This was carried at the upper end of an upright shaft which ran in footstep bearings mounted on horizontal timbers or sprattle beams. At the lower end of the shaft is a large spur wheel which engages with the stone nuts. The stone nuts may be carried on quants driving the stones from above or, alternatively, mounted on the stone spindles and driven from below. A stone driven from above is known as an 'overdrift' stone, the alternative being 'underdrift'. Both overdrift and underdrift runner stones are similarly mounted on the squared end of the stone spindle by the means of a mating socket in the mace, but the spindle of the underdrift stone, in addition to bearing the stones weight, also transmits the drive. To disengage the drive the stone nut is arranged to slide axially on the stone spindle, the actual movement of the nut being carried out by means of a screw, forked lever or something similar, the method adopted varying from mill to mill.

Cutaway view of a pair of overdrift mill stones 'exploded' to show:
1 Stonenut
2 Quant
3 Runner Stone
4 Gymbal Bar
5 Mace
6 Bedstone
7 Stone Spindle
8 Bridging Box

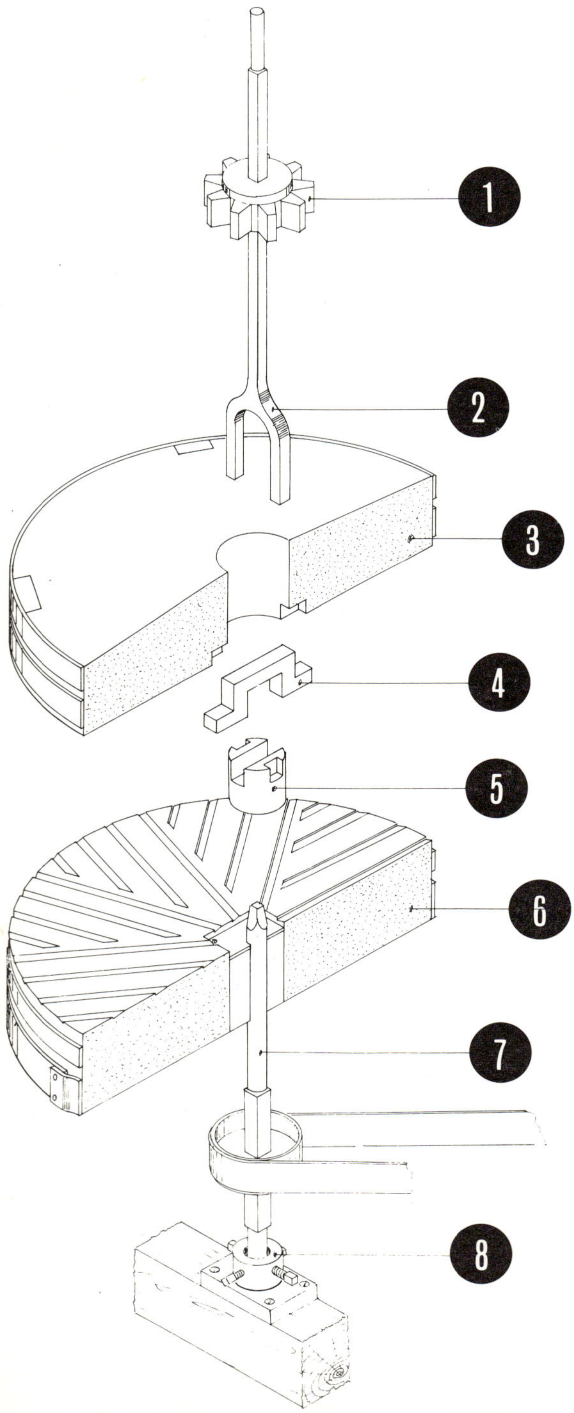

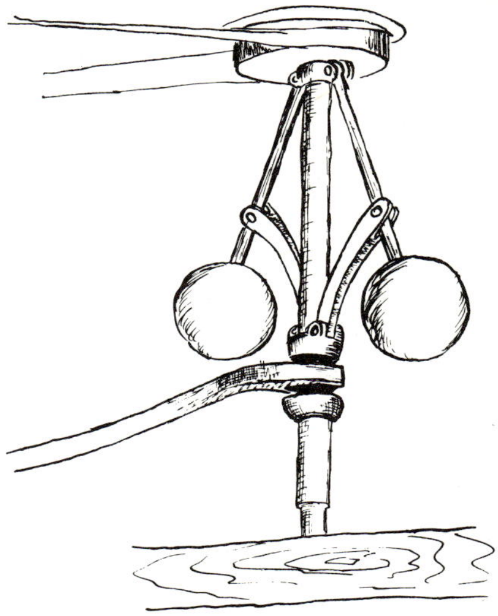
Governor: Bourn Mill

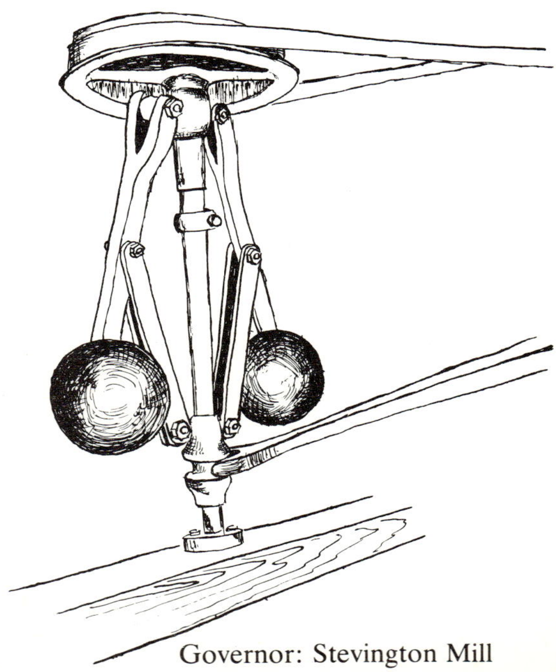
Governor: Stevington Mill

This diagram shows the layout of the drive to a pair of underdrift stones. A second pair of stones is shown beside the first, an arrangement common in post mills. Three and sometimes four pairs of stones arranged around the great spur wheel are met with in the larger smock and tower mills. The numbered parts are:

1 Damsels
2 Runner Stones
3 Stonenut
4 Stone Spindle
5 Wallower
6 Upright Shaft
7 Great Spur Wheel

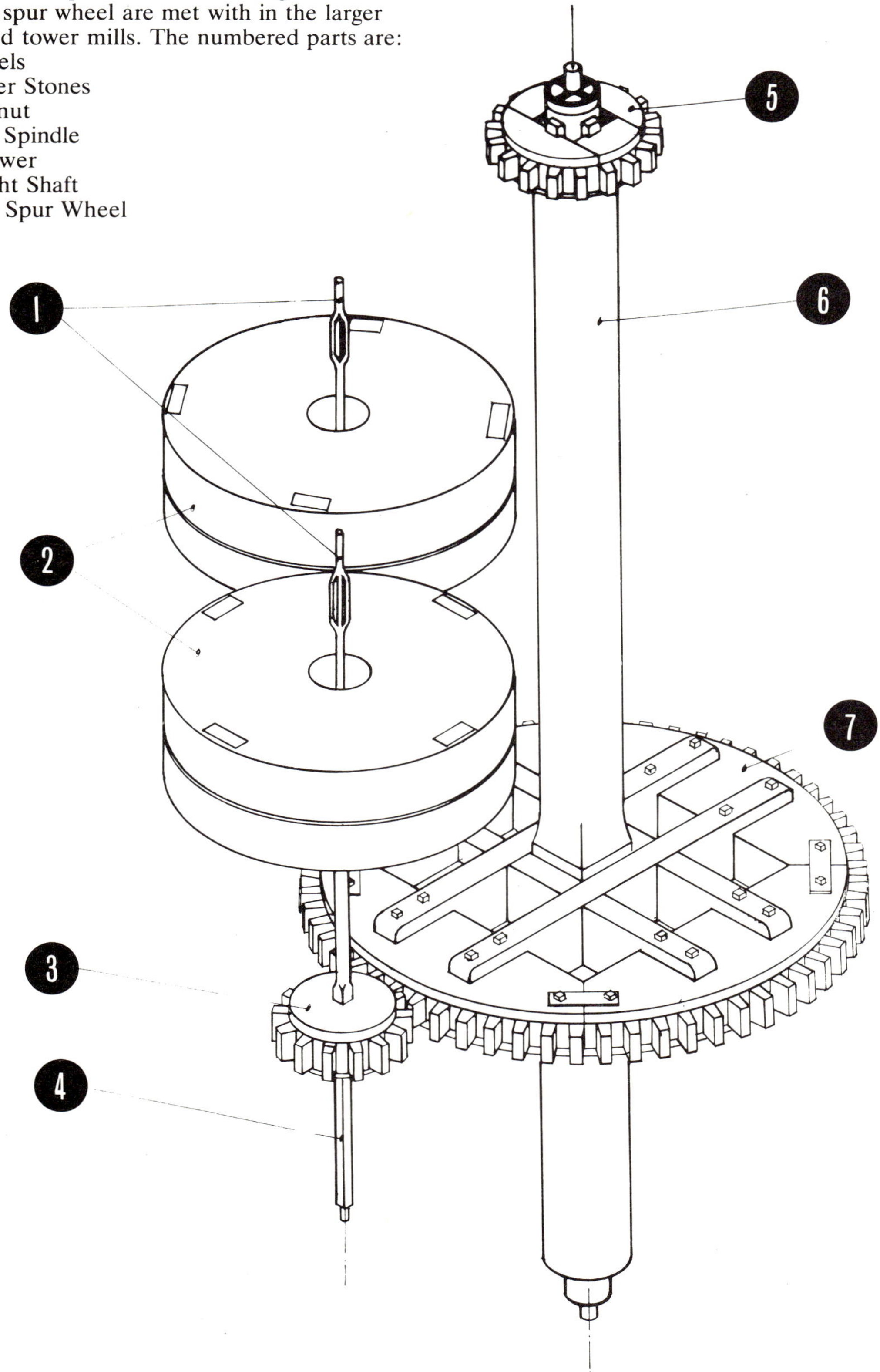

Grain feed to the stones—rough sketch showing grain feed:
1 Meal Bin
2 Meal Spout
3 Bedstone
4 Runner Stone
5 The Eye of the Stone
6 Crook String
7 Hopper
8 Horse
9 Shoe
10 Stonevat

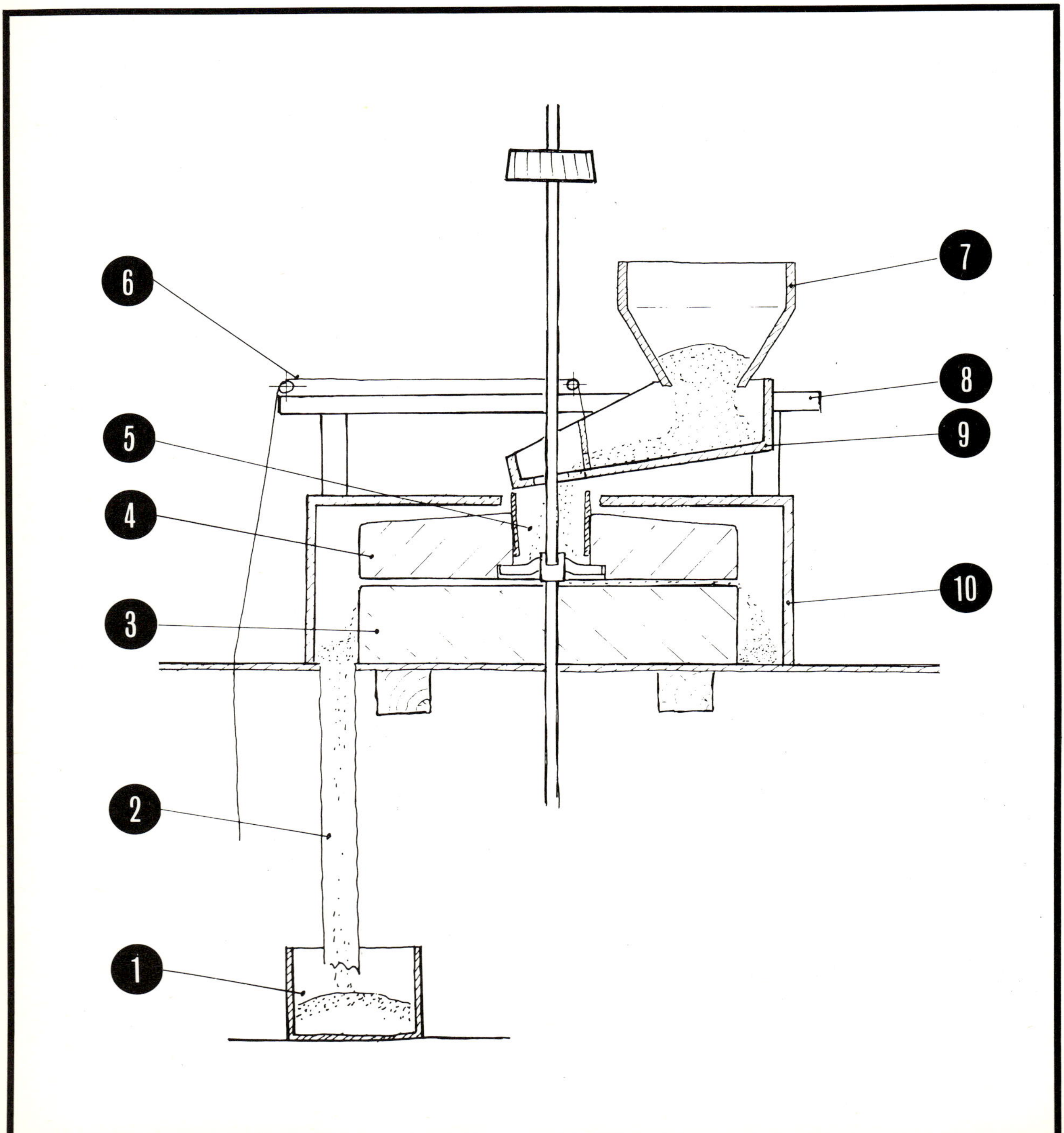

Diagram of a Typical Tentering Gear showing:
1 Quant
2 Runner Stone
3 Bedstone
4 Stone Spindle
5 Bridging Box
6 Governor
7 Steelyard Loops
8 Steelyard
9 Adjusting Screw
10 Bridge-tree
11 Brayer

In operation if the speed of the runner stone increases, the weights of the centrifugal governor move outwards and the links attached to the arms draw the sliding collar up the governor spindle taking the forked end of the steelyard up with it. The outer end of the steelyard will therefore be pivoted downwards thus lowering the brayer and in turn the bridge-tree and runner stone.

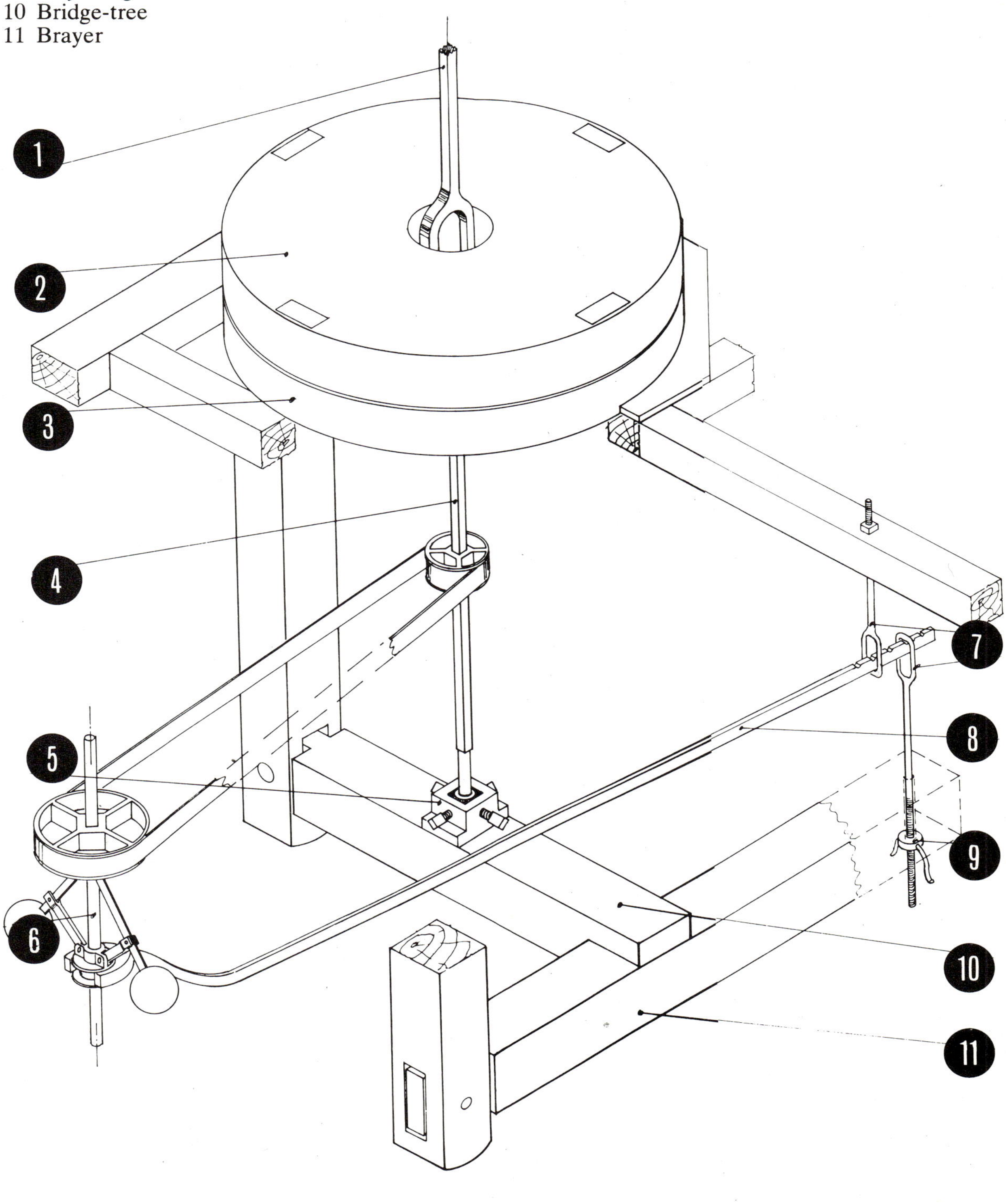

The Grain Feed to the Stones

The grain was usually carried in bins on the floor immediately above the stone floor and gravity fed to the hoppers through wooden chutes. Each hopper was carried on a wooden framework known as the horse, mounted on top of the stone vats, the grain being fed to the eye of the runner stone by way of an inclined wooden trough or shoe held against a square section on the rotating quant by means of a leaf spring. As the quant rotated, the shoe was shaken and the grain was kept moving down the incline. In the instance of an underdrift stone a multi-armed spindle, known as the damsel, due it is said to the noise it creates, protruded above the runner stone and agitated the shoe in a similar fashion. The inclination of the shoe could be adjusted from the spout floor below by the crook string which ran over a pulley on the horse, then passed through a hole in the floor and was wound onto the 'twist peg', a simple but effective means of regulating the volume of grain fed to the stones. From the eye of the runner stone the grain passed between the stones, was ground and issued from the circumference of the stones as meal. A projecting tag on the outer circumference of the runner stone swept the meal around the bottom of the vat until it fell through a hole in the bottom and thence, via a wooden chute or stocking like spout, into bins or sacks on the floor below. To avoid the danger of the stones being starved of grain and touching, an alarm bell was invariably rigged up by the miller to indicate when the hoppers required re-filling. Contact of the stones would dull the cutting edges of the furrows or worse might create sparks and fire the mill. The sacks of grain and meal were raised and lowered by means of the sack hoist, situated in the top of the mill, each floor having upward opening trap doors to facilitate the passage of the sacks up through the mill.

Tentering

The degree of fineness of the meal was dependant on the size of the gap between the stones. This gap however had to be regulated to suit the speed of the sails, which was of course dependant on the strength of the wind. For example, if a fine quality meal was required the stones would be set very close, in which case a very strong wind was needed to overcome the increased friction set up by the passage of the grain between the stones. This could cause a problem because if the wind strength lessened, the stones might well cease turning. On the other hand, if the stones were set well apart to suit a light wind (thus producing a coarse meal) and the wind strength increased, the stones would race with the danger that the mill might 'run away'. The stone gap therefore required frequent adjustment to suit the varying wind strength, and on the earliest mills the operation, which was known as 'tentering', was carried out manually, and in gusty winds must have been an irksome task. A means of regulating the gap automatically by the action of a centrifugal governor was eventually devised and is another instance of the extraordinary ingenuity of the early millwrights.

As mentioned before, the lower end of the stone spindle runs in a bridging box on the bridge-tree, a stout beam pivoting at one end, but free to be raised or lowered at the other. Raising or lowering the bridge-tree therefore will also raise and lower the runner stone. The free end of the bridgetree is suspended on the end of another pivoting beam known as the brayer, which in turn has its free end supported by the steelyard. The steelyard pivots are at a point close to the end supporting the brayer, thus a tremendous mechanical advantage is obtained, and a small effort applied to the outer end of the steelyard being all that is required to raise a runner stone weighing perhaps a ton. This 'small effort' is obtained from a centrifugal governor usually driven by a belt and pulley from the stone spindle or perhaps the upright shaft. The outer end of the steelyard is forked and locates in a grooved collar on the governor shaft. This collar slides up and down the governor shaft as the governor responds to changing wind strength and, via the steelyard and brayer, raises or lowers the bridge-tree, thus altering the stone gap in relation to the strength of the wind. The manual screw of the hand tentering gear is retained to provide a means of initial setting, but once set by the miller the automatic tentering gear enabled grinding to continue in gusty winds without frequent interruptions for adjustment.

Flour Dressing Machines

The meal in the form it leaves the stones consists of a mixture of flour and bran which has to be separated. In the earlier mills this was done by hand with a sieve, but by the early part of the 18th century machines to carry out the sifting were being increasingly fitted into both new and existing mills.

The Jog-Scry

One of the earliest and simplest forms of flour grader was the jog-scry, nowadays seldom met, but a few examples have survived. It consisted basically of an inclined chute made up of a number of wire sieves which were in increasing degrees of coarseness towards the lower end. One end of the chute was hinged and the other was suspended on a belt which was joggled by the action of a cam or small crank, the method varying from mill to mill. The agitation imparted to the chute shook the meal down the incline, grading it into various qualities.

The Bolter

The jog-scry was superseded by the bolter, a more sophisticated machine to carry out the same process. It consisted of a wooden cylindrical reel which rotated on an inclined axis and was encased in a rectangular wooden box. The reel was covered with a stocking-like woollen bolting cloth which had a mesh, varying from fine at the top end of the reel to coarse at the lower end. The reel was open at both ends but the upper end was protected by a coarse wire mesh cover to keep out foreign bodies. Meal was fed into the upper end of the rotating reel from a hopper above and the weight of the meal as it progressed through the reel made the cloth sag and knock against wooden bars running the length of the casing and spaced around the reel. The intermittent striking of the cloth against the bars shook the flour through the cloth, the finest grade through the top mesh and into a chute, the slightly coarser through the next and so on, leaving the bran and middlings to escape out of the lower end. Bolters were very effective machines but suffered from one serious weakness, the bolting cloth, which not only wore out rapidly due to the hard life it led, but also suffered depredation by moth.

The Wire Machine

Similar in principle to the bolter, but instead of a rotating reel, it had a spindle carrying a number of radial brushes. This was mounted in an inclined cylindrical housing lined with sections of wire mesh in varying degrees of fineness. In operation the rotating brushes swept the flour through the wire mesh, grading it into several qualities of fineness, which descended to the floor below through stocking-like spouts feeding the various bins. A bolter or wire machine was usually carried on the meal floor or, in many instances, the stone floor, which in a smock or tower mill was no problem, there being plenty of space. Fitting these machines into the smaller post mills however was rather more difficult, the most satisfactory solution being to add an extension at the tail of the body. Examples of post mills with such extensions may still be seen, for example, the Cat and Fiddle Mill in Derbyshire and Great Chishill in Cambridgeshire. Both bolters and wire machines were usually belt-driven from a countershaft, the initial drive to which in a post mill often being from the brake or tail-wheel via a skew gear. In smock and tower mills the countershaft was commonly driven either from a bevel gear on the upright shaft or a bevel ring on the great spur wheel. Other alternative arrangements were contrived of varying complexity and ingenuity but all had one point in common, they worked!

'Jack' and 'Jill' windmills at Clayton, Sussex. A rare survival of a smock and a post mill standing alongside each other. (Photo: A.F. Kersting).

Standing in 'solitary isolation' a bare, stark sentinel of a useful past. The sad remains of Ringmer Mill—the centre post was 20-ft. high and 2-ft. thick. Now, even this has gone.

Types of Windmill

Whilst much of the machinery of windmills so far described is basically similar for both post and tower mills, structurally they require separate description.

The Post Mill

The post mills which survive in England today are very similar to the earliest type of mill shown on medieval manuscripts. Refinements such as patent sails and fantails were added in the course of time, but the principle on which the post mill was based remained unchanged until the end of the windmill era.

The rectangular body or buck of the mill containing all the machinery is pivoted on a vertical post to enable the sails to be brought into the wind. The sub-structure or trestle which holds the post upright was on the earliest mills sunk into the ground to give stability, a practice which caused rapid deterioration of the timbers. The trestles of the later mills were raised clear of the ground on brick or stone plinths and eventually, to provide much needed storage space and protect the sub-structure from the elements, roundhouses were added. Many open trestle mills were modified in this way and most had to be raised onto higher piers to obtain a useable space beneath the trestle; this often provided an additional advantage, for by raising the mill clear of obstructions, such as nearby trees or buildings, better wind conditions would be obtained at the sails.

The body of a post mill is constructed entirely of wood and has two or sometimes three floors. Working down from the top, these are the bin floor carrying the grain bins; the stone floor where the actual grinding is done; and the spout floor where the meal spouts and bins are located. Chutes from the bin floor kept the hoppers on the stone floor supplied with grain, but often mills would have no bin floor, the grain being loaded into the hoppers direct from sacks. Access to the body is by the wide ladder leading to a door at the rear of the spout floor. The roundhouse below may be built of brick, masonry or timber, and have one, two or even three floors. The roof of the roundhouse is usually conical in shape and open at the apex immediately below the spout floor, the ingress of rain being prevented by the overhanging weatherboarded sides of the body. No access is normally provided into the body from the roundhouse except for the trap door in the spout floor which enabled sacks of grain to be raised by means of the sack hoist. This would be situated in the apex of the body and consisted of a simple wooden bollard usually driven by belt or chain from a pulley on the windshaft.

The post on which the body pivots is usually oak of considerable girth, about two feet six inches square being usual, and is held upright by four diagonal struts or quarterbars mortised into it. The lower ends of the quarterbars are likewise mortised into the outer ends of two heavy horizontal beams or crosstrees, laid crosswise below the post, the outer ends of the crosstrees being supported on brick piers. The lower end of the post is quartered over the junction of the crosstrees to provide a location, but the crosstrees do not bear its weight, an appreciable gap being present at this point to ensure that the weight of the mill is transferred from the post and quarterbars to the brick piers which, therefore, take the whole weight of the mill. The crosstrees act as tie beams, and in conjunction with post and quarterbars, form a triangulated structure of great strength.

The upper part of the post which projects above the roundhouse into the mill body is usually round or chamfered in section and capped with a pintle. On the very early mills this was turned from the solid oak and it located in a socket cut into the underside of a massive oaken beam known as the crowntree which pivots horizontally on the post. On the later post mills iron pintles were used in conjunction with a samson head, a casting fixed to the crowntree. The crowntree is midway up the mill body, immediately below the stone floor and runs the full width. Fixed at each end of the crowntree are stout horizontal members known as the side girts, which run the length of the body and support the corner posts. On this basic structure the very rigid mill body framing is built, held together by mortise and tenon joints pinned with hardwood dowels, and on many mills iron tie rods, in most cases later additions, lend extra strength to the structure. The body framework was covered with horizontal weatherboarding either creosoted, tarred or painted white, removeable hatches being provided in the sides to light the interior.

To keep out driving rain at the corner joints, the weatherboarding on the breast projected two or three inches beyond the sides and similarly that on the roof and sides overlaps the boarding at the rear. A mill being usually set facing the wind and thus also driving rain, this overlap provided an effective means of weatherproofing. The roof of the earliest post mills was a simple straight pitched affair, (a single example of which has survived at Bourn), but this later developed into the curved shape now familiar, and enabled the large brake and tail-wheel to be accommodated without wasting space.

Immediately below the lowest floor of the mill are two heavy section beams or sheers running fore and aft on either side of the post. Between these sheers and around the post is a wooden bearing or collar to steady the mill as it pivots. Protruding from between the sheers at the rear of the mill and passing through the ladder, is the tailpole; this is angled towards the ground and of sufficient length to provide the leverage necessary to turn the mill. At the end of the tailpole is a lever known as the talthur, which is connected to the ladder by a chain and is used to raise the ladder, which is hinged at the top, clear of the ground when turning the mill to face the wind.

Fixed between the fore-end of the sheers and running vertically up the front of the body is the prick post, a timber providing support to the weather-beam, a stout horizontal timber which carries the neck bearing of the windshaft. The socket bearing for the rear end of the windshaft is carried on the tail-beam, which lies horizontally across the rear of the body but set slightly below the level of the weather-beam to give an inclination of about 10–15 degrees to the windshaft. The purpose of this inclination is to transfer a proportion of the great weight of the sails and windshaft assembly to the tail bearing, it also enables the sails to clear the roundhouse. Mounted on the windshaft just behind the neck bearing is the brake wheel and at the other end is sometimes carried the similar but usually smaller tail-wheel.

In the older mills the drive to the stones was direct from the brake and tail-wheels and this dictated the stone layout known as 'head and tail'. This arrangement provided good weight distribution fore and aft of the crowntree and hence fine balance on the post, essential both for ease of turning the mill into wind and ensuring the correct horizontal setting of the runner stones in relation to the bedstones. Later post mills were built with both pairs of stones in the breast, a position which necessitated an indirect drive via an intermediate upright shaft. With this arrangement the stones were as often as not driven from below and sometimes a third pair of stones were mounted in the tail, either driven direct from the tail-wheel if present, or underdrift from the great spur wheel on the upright shaft.

To obtain maximum benefit from the wind, the sails must face the quarter from which it is blowing, and although winding a post mill by means of the tailpole was, due to its fine balance, a fairly easy task, a means of keeping the sails facing square to the wind automatically was required. The fantail provides this means. When the fan is rotating it drives, through a system of bevelled reduction gears, a wheeled carriage at the end of the tailpole or ladder, which runs on a firm track. The action of the fan will thus pivot the mill body on the post bringing the sails round to face the right quarter. When the carriage is mounted on the end of the tailpole similar wheels and track are provided for the ladder to allow it to rotate with the mill body. The alternative to tailpole mounting is for the fan carriage to be integral with the ladder, in which case the super-structure carrying the fan is quite high, both to allow free passage underneath and to catch the wind—this type of fan tackle was very common in Suffolk. A third system is for the fan to be mounted on the rear of the body roof, an effective position for catching the wind. This fan position did however entail a lengthy system of shafting to the wheels at the foot of the ladder, or less commonly to gears meshing with cogs on a circular rack mounted on top of the quarterbars.

Although the post mill could not compete with the later smock and tower mills in terms of meal production, its constructional cost was much lower and the building of post mills continued well into the 19th century. Today many post mills are still to be seen, a tribute to both the skill of their builders and the enduring qualities of English oak.

The post mill on Outwood Common, Surrey, in 1927. This mill has now been expertly restored, as our colour picture on page 126 shows.

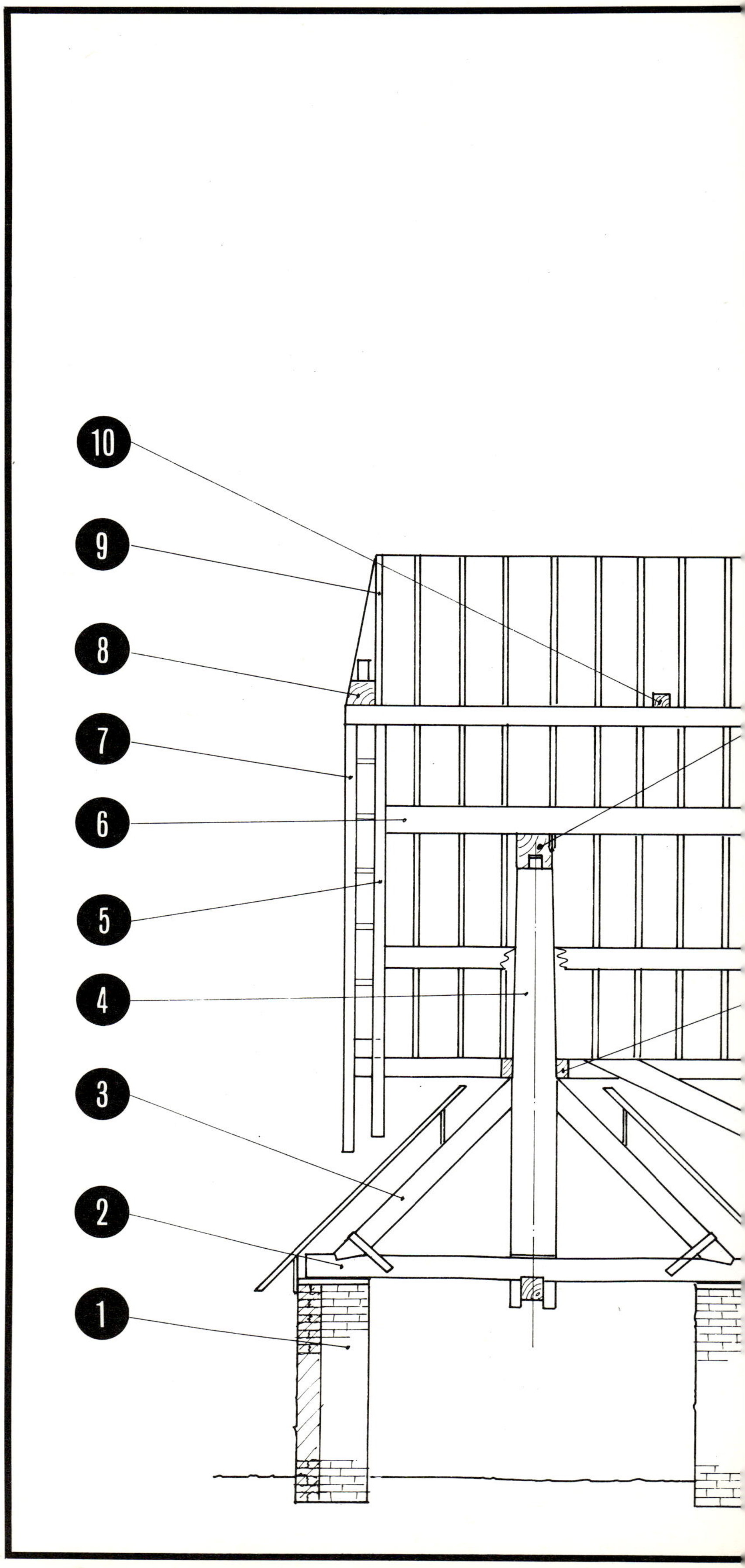

Diagram of Post Mill Structure:
1 Brick Pier
2 Crosstree
3 Quarterbar
4 Post
5 Corner Post
6 Side Girt
7 Prick Post
8 Weather-beam
9 Roof Rib
10 Tail-beam
11 Collar
12 Sheer
13 Tailpole
14 Ladder
15 Talthur
16 Crowntree

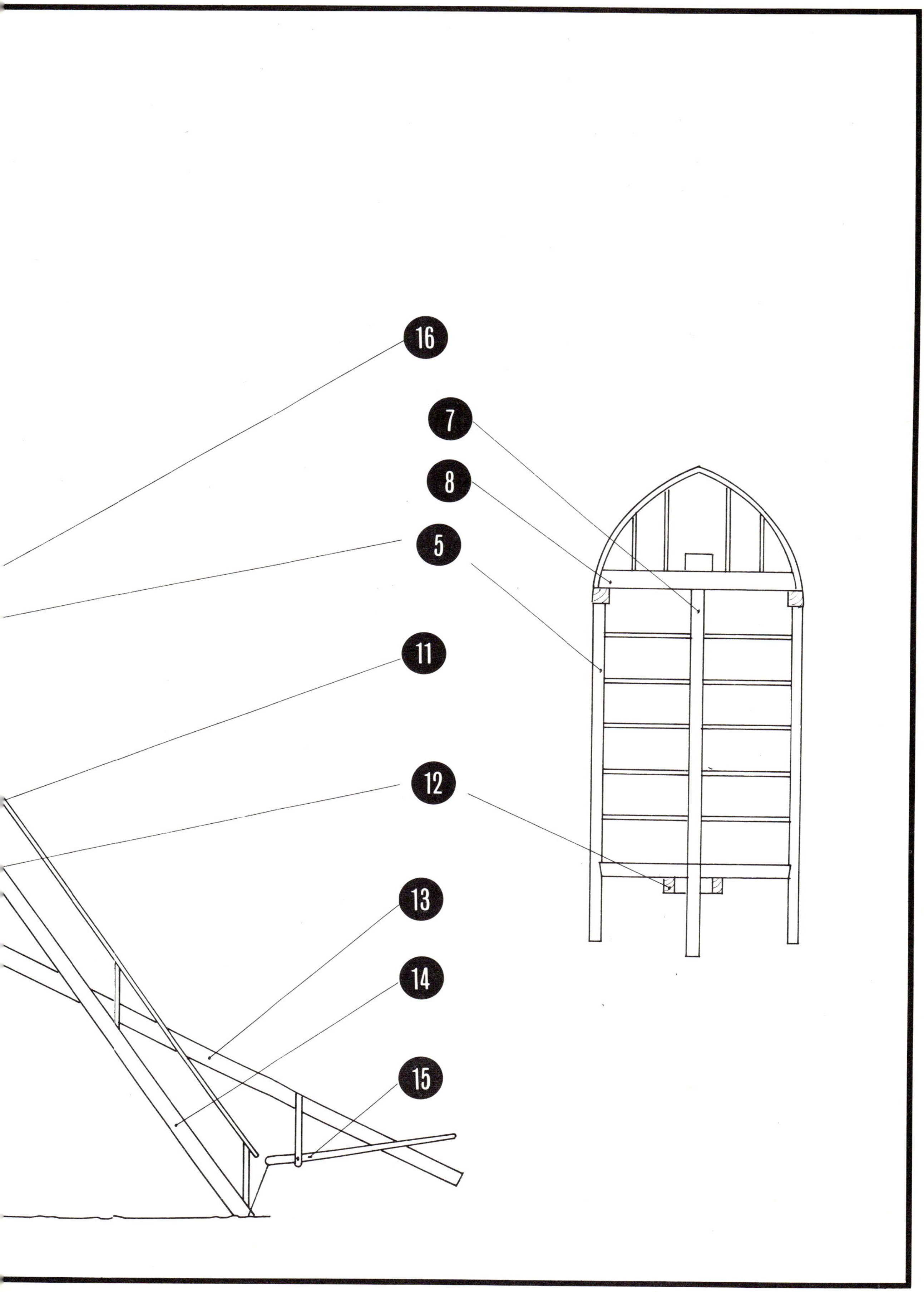
16
7
8
5
11
12
13
14
15

Smock and Tower Mills

The fact that the entire body of a post mill had to be turned to bring the sails into the wind imposed a limit on the size of this type of mill. Later types which overcame this drawback were the smock and tower mills. In these, the body containing the stones and machinery is fixed and has a relatively small rotating cap or cupola mounted at the top. The cap carries the windshaft and sails and can be turned to face the wind either manually or by means of a fantail, unhampered by the weight of the machinery within the mill body. Because of the design of the cap and the manner in which it functions, the drive to the machinery below is transmitted via a centrally located upright shaft. From the great spur wheel mounted lower down on the upright shaft the final drive to the stones is similar to that in a post mill and may be underdrift or overdrift. The internal economy of smock and tower mills is similar, they differ only in the choice of the materials used in the construction of the body, the smock being built largely of timber and the tower of brick or masonry.

Smock Mills

Smock mills were probably introduced in the 17th century and are said to have acquired their name due to a supposed similarity to the former countryman's working smock. Usually octagonal in plan, although other shapes are met with, the body is mounted on a brick base which may consist of just a few courses or be up to three storeys high. The structural framework is composed of stout corner or cant posts, usually oak, braced together and inwardly inclined, which are tied at the top by a circular track or curb. At each floor level horizontal tie beams are mortised into the cant posts and these support cross beams which carry the floors. This very rigid framework is covered with horizontal weatherboarding, the butt joints overlapping at the cant posts and covered with zinc or lead strip to keep out rain. The weatherboarding is either tarred black or alternatively painted white, the latter an attractive but expensive finish to maintain. Whereas post mills had removable hatches to light and ventilate the interior, smock and tower mills had glazed windows and were therefore much more comfortable to work in under weather conditions that were so often the millers' lot. A weakness of the smock mill lay in the vulnerability of the structure at the corner joints to the ingress of rain and subsequently rotting of the timber. The resulting distortion of the framework of the mill could cause buckling of the curb and hence an inoperable cap.

Haverhill Mill, a tower mill, the first to be built with the annular or circular sails. This rare photograph was taken in 1929 when the mill was working and up for sale. Alas, this mill has since been demolished.

Tower Mills

Tower mills are constructed of brick, masonry or similar durable material and were the ultimate phase of windmill development, many being built which were over ninety feet high and carrying seven or eight pairs of stones. Normally round in plan, although polygonal shapes were not uncommon, the body is usually tapered like a smock mill and for the same reason—strength and to reduce wind resistance immediately behind the sails. A covering of tar or cement rendering was often used to weatherproof the tower or less commonly the brickwork was sometimes tiled. In the better designed mill the windows were arranged in a spiral fashion around the tower to avoid subsidence of the brickwork and distortion of the curb, a common occurence when window or door openings were set one above the other.

Caps of both smock and tower mills are made of timber, the basic structural framework consisting of two stout horizontal sheers running fore and aft, which often protruded at the rear to provide a mounting for a fantail. The fore-ends of the sheers are tied together by the weather-beam and at the rear by the tail-beam; these carry the neck and tail bearings for the windshaft in a similar fashion to the arrangement in a post mill. Fixed to the sheers midway between the weather and tail-beams is a sprattle beam, a substantial timber which carried the upper bearing for the upright shaft. A similar timber and bearing support the lower end of the shaft. These bearings are positioned at the axis of the cap to ensure correct meshing of the brake wheel and wallower irrespective of which way the cap is facing. The cap framing is built up on this basic structure and covered with a suitable weather resistant material, such as tarred or painted canvas over wooden boarding, lead or copper sheeting, or even thatch.

Mounted on the underside of the cap and running on the curb are rollers or, alternatively, well greased brass skid plates. When rollers are used the curb is known as a live curb; with skid plates it is a dead curb. To keep the cap central and prevent it overriding the curb, a set of wheels mounted on vertical axles run beneath a flange on the curb's inner circumference. Curbs are normally made up of timber segments with an iron facing and carry a toothed rack around the circumference. A pinion or worm wheel meshing with the rack, takes the drive from the fantail, or with manual winding of the cap, an endless chain and pulley which via suitable reduction gearing rotates the cap to face the wind. In many cases an outside gallery was provided to facilitate routine maintenance to the fantail and driving mechanism, a task only for those with a head for heights! In addition to the purpose of bearing the sails into the wind, the cap had another function, namely to protect the curb and upper machinery from the elements. The weather-beam and neck bearing of the windshaft being particularly vulnerable to driving rain and snow, a deep vertical board was sometimes fitted at the front of the cap to act as a shield, or in the case of a circular cap a deep petticoat, often attractively scalloped, overhung the curb.

Watts' Smock Mill, Hildenborough, Kent, 1930

The shape of the cap on a mill tower without doubt contributes more towards the attractiveness of the mill than any other of its features. Several distinct shapes are met with, some of which predominated in certain parts of the country. Boarded caps shaped like a post mill were common in Essex and Kent, whereas East Anglia favoured caps not unlike an upturned boat, whilst Lincolnshire is noted for its elegant ogee caps topped with either a ball or acorn finial, but as the craft of millwrighting spread, mills with features hitherto confined to a particular locality made their appearance in other areas.

A typical smock or tower mill would have five floors. These from the cap downwards are: the dust floor which prevents dust from rising into the cap, and protects the lower floors from dirt falling from above. Immediately beneath is the bin floor where the grain is carried in bins which feed the hoppers on the stone floor via chutes. The stone floor where the grinding is done is immediately below the bin floor, and below this is the spout floor where the meal is fed into sacks. The spout floor, or meal floor as it is sometimes called, normally carries a bolter and perhaps other machines as well, all driven by variously contrived gearing or pulley and belt systems. The ground floor is normally used for the storage of grain until needed on the bin floor to feed the hoppers and in many mills might also contain an auxiliary engine to power further stones which were brought into use on windless days. As in the post mill, upward opening trap doors were provided in each floor to allow the sacks of grain and meal to be raised and lowered by means of the hoist which is usually situated on the dust floor. A typical sack hoist would consist of a wooden bollard driven by a friction wheel in contact with the underside of the wallower or, alternatively, by a belt and pulley from a countershaft, this being driven by gearing from the upright shaft. An outside stage was often built around the mill body at spout floor level from which the sails could be adjusted, access to the stage being provided by opposite doors, the one used being the one furthest from the moving sails.

Although superior in terms of meal production, the smock and tower mills did not entirely supplant the post mill. The quality of the meal produced in a post mill was equal to that from the later mills, the actual grinding process being the same. Post mills were certainly less costly to build, a factor which often made them a more attractive proposition on an untried site. Indeed, most of the larger smock and tower mills were built on sites, the suitability of which had been proved by the successful working of an earlier post mill. It was not uncommon for a smock or tower mill to be built and worked alongside the post mill it was to eventually replace, although in some instances the newcomer was outlived by the older mill, as at Outwood.

Drainage Mills

Derelict mill towers are a familiar sight in the marshlands of the Fens and on the banks of the rivers of the Norfolk and Suffolk broads. These are the remains of the drainage mills, the forerunners of which were brought into use in the 16th century to help in the tremendous task of the reclamation of valuable tracts of flooded land. Dutch in origin, the first drainage mills were simple smock mills fitted with common sails and winded by tailpole. Development kept pace with its corn grinding equivalent and the later drainage mills, of which there were many hundreds, were brick built towers, frequently well tarred, carrying patent sails and winded automatically by fantails.

A typical drainage mill would be sited beside a river, into which water drawn from the adjacent marshland could be discharged. Its internal mechanism is rudimentary, consisting basically of a pair of bevel gears at the top and bottom of the tower, connected by a centrally located upright shaft. Driven from the lower end of the upright shaft is a horizontal axle which protrudes through the side of the tower and carries the scoop wheel. The average scoop wheel will measure between twelve and twenty feet in diameter and be housed in a semi-circular wooden casing which is usually built against the side of the mill tower. The lower half of the scoop wheel is submerged in a culvert the width of which closely approximates to. that of the wheel. In operation the scoop wheel carries the water up out of the culvert and discharges it at a higher level through an aperture into the river. The aperture is fitted with a self closing hinged flap to prevent water flooding back in the event of the river level becoming too high. The bevels at the upper end of the upright shaft are similar to those in a corn mill, the one on the windshaft being the brake wheel and the other the wallower.

On the ground floor, the driven bevel of the pair is mounted on a common axle with the scoop wheel and is half submerged in a pit, hence its name, the pit wheel. All the intermediate floors between ground level and the dust floor are empty of machinery, except for the upright shaft which passes through a hole in the centre of each floor. In a few cases drainage mills were fitted with millstones for corn grinding and in at least one instance (described in a later chapter) cement clinker. Such dual purpose mills were however exceptional to the general practice. The decline of the drainage mill paralleled that of the corn-mill and it was eventually overtaken and ousted by pumps, at first powered by oil engines and later electricity.

The Hollow Post Mill

Unlike the normal post mill, the body of this type of mill contained no machinery except for the windshaft with its attendant brake wheel and wallower and was therefore in consequence much smaller. The post on which the body pivots was hollow and the upright shaft was carried down through it into the roundhouse or lower compartment of the mill to drive the machinery.

Originating in Holland, where it was widely used for drainage, and dating from the 15th century, the hollow post mill was not built in great numbers in England and, unfortunately, no complete example has survived.

Composite Mills

A by no means uncommon failing with the post mill was a defective trestle, due to either deterioration of the timbers or subsidence of one or more piers. Either fault could put the mill out of action due to the difficulty of maintaining the stones absolutely level. Many mills did of course have parts of their sub-structure renewed. The alternative, in a few instances, was to mount the sound body on a short brick tower built especially to receive it. The resulting hybrid is known as a composite mill. The underside of the body was fitted with rollers which ran on a curb in the manner of a smock or tower mill cap, winding being in some cases by tailpole, in others by fantail. Few mills were converted in this fashion however, probably because it was found that the replacement of the trestle with a tower offered no advantages either in cost or efficiency over the fitting of a new trestle.

Leigh Green Mill, Nr. Tenterden, Kent

Barham Mill, Kent. Built *c.*1834. Totally destroyed while repair work was being carried out in 1970 due, it is thought, to a stray spark from the workmen's bonfire.

Fire, Storm and Tempest!

What a wealth of windmills would remain but for the ravages of fire, storm and tempest, to which many have succumbed. In just one great storm in 1703, four hundred of these magnificent monuments were lost, and every year since more have been doomed. It is only in recent times that this tide of destruction has been stemmed and some concerted efforts have been made to preserve these remnants from a bygone age.

Mills were, unfortunately, a fire risk—the dust laden ledges forming a ready kindling to a stray spark. Many stood in high places to catch the merest breath of wind, and their very situation made it difficult to quell the flames, since water was not always close at hand.

Little wonder that the shout of 'Fire!' struck fear into many a merry miller's heart.

Fire !

'When I first arrived on the scene dense clouds of smoke and a little flame were issuing from out of the top of the structure but gradually the dark red fire showed itself......
The glass was shattered in the little windows and then the draught was increased. The roar of the fire grew louder and louder, and the heat became more and more unbearable, steadily driving the crowd backwards. The doors soon opened and then the scene was a splendid one, the interior of the mill could be seen to be a veritable furnace, nothing but a mass of flame was visible. Here and there a tiny jet of light and small curling wreaths of smoke crept from between the woodwork, and the wind fanning the fire, in a very few minutes the gallery and other parts were seen to be well ablaze. The flames crept up and caught the sails, and then the end approached. The whole was almost completely enveloped in a wild mantle of fire. The heat thrown out was tremendous and, at times, the black smoke clouds were overpowering......Then the crash! Down came the whole concern.
It did not sway at all, but gave one the impression of a huge concertina being compressed. At the moment of the fall a 'heatwave' of fearful power was caused and everyone was compelled to retire in haste covering his face. The adjoining buildings, which had also been steadily burning and which were well tarred, collapsed. The lower part of the mill, of brick, was simply a seething furnace, the machinery being white hot. Rats and mice crawled about and ran hither and thither in their flight'.

Lydd Mill, Kent. Originally built in 1805, this smock mill succumbed to the flames on 22 September 1927, as this photograph shows.

This graphic account of the destruction of the White Mill at Hastings appeared in the *Hastings and St. Leonards Observer* on the 19th May 1900. Windmills being predominently wooden structures, and frequently liberally tarred to resist the elements, were particularly vulnerable to fire which once started would quickly take hold and sweep through the mill. The fact that many windmills were in extremely isolated situations usually ensured, in the event of a fire, that by the time the firemen arrived on the scene there was nothing left worth saving. The ways in which a fire could start were many and varied; oil lamps and candles used to illuminate the interior at night were the main hazard. Another source of danger was the millstones; if the flow of grain from the hopper to the stones was cut off, the stones could touch and create sparks which in the dry and dust laden interior of a mill could easily cause a blaze. To avoid this danger, simple but ingenious bell alarms were invariably fitted to grain hoppers by the millers to ensure that the millstones were not allowed to be starved of grain. A sudden increase in wind strength could be a potential fire hazard, especially before the introduction of self-reefing sails, because the friction caused by the application of the brake to arrest a fast running mill could be sufficient to start a fire and a miller who was taken unawares by a sudden squall had perforce to quarter the mill, (set the sails at right angles to the wind), and at the same time feed the stones with as much grain as possible in an effort to slow them down. In some instances, in all probability a fire was a convenient means of getting rid of an unwanted mill, one which perhaps needed costly repairs which the owner considered uneconomic. In such cases the difficulties of demolishing the structure with its attendant high cost could be avoided by an 'accidental' fire.

The tragic loss of the fine smock mill at Barham in Kent, as recently as 1970, emphasizes the need for care when lighting fires near timber buildings. On the 3rd March in that year workmen arrived at the 136 year old mill to begin repairs. Shortly after lunch flames were seen in the mill which, encouraged by a moderate breeze, quickly spread through the structure and, being well tarred, it soon became a raging inferno. Although strenuous efforts were made by the local fire brigade to quell the blaze, two firemen actually entering the mill, it was to no avail, and the villagers who had been trying to raise funds for restoration work could only look on helplessly as the mill was completely destroyed. The fire was thought to have been caused by a stray spark from the workmen's bonfire, lit to burn rubbish. Barham Mill, which incidently had the distinction of being chosen as the setting for the Kenneth More film 'Raising a Riot', was a well-known and attractive landmark; it had been conscientiously maintained in working order for many years; new sails had been fitted in 1962, and its cap had been repaired only a few months previous to the fire.

The Barham Mill fire evokes memories of another sad loss during the fifties, when the old post mill at Sprowston in Norfolk was burnt down. This mill was supposed to have inspired John Crome's famous painting *A Mill on a Heath near Norwich* and which now hangs in the National Gallery. There is no real evidence to support this story however. The mill had not been out of use for many years before its destruction and because of its historical associations, a fund had been set up to acquire and maintain it. Ironically this had been almost finalised when the mill was destroyed. A very fine working model of the mill is displayed in the Science Museum at South Kensington.

The modern practice of stubble burning is often negligently carried out; in at least one instance a pair of thatched cottages being completely destroyed. It was disquieting to note during a recent visit to the nicely restored post mill at Pitstone that the cornfield in which the mill stands had been cleared by stubble burning. The likelihood of this mill, which took six years to renovate, surviving for many more years if this practice continues would appear slight.

Lydd Mill after the fire

Above: Rye Mill in 1920. This famous Sussex landmark was severely damaged by fire in 1930.

Below: The sails of Wangford Mill, Suffolk, crashing to the ground during a fire in 1928, which destroyed the well-known landmark. The blaze was visible for many miles around and attracted many helpers and spectators.

Barham Mill, before the fire.

Below: The historic Sprowston post mill on Mousehold Heath, Norwich (made famous by John Crome, the artist, and George Borrow), before being destroyed by fire.

Left: The mill at Meopham Green, Kent, that was saved during a gale in 1935 by the gallant efforts of the owner and his two sons. In recent years, this mill has been restored and the missing sweeps replaced.

Below left: Firemen at work amid the ruins of Sprowston mill, Mousehold Heath, Norfolk.

Below: Cloke's mill, Waltham, Kent. This tarred smock mill fell in a gale in February 1931. Its interior of beams and metal workings were thrown far across the meadow.

Bluebell Hill mill in Kent as destroyed in a gale July 1890.

Upchurch mill, Kent. Burnt down September 1910.

Bekesbourne mill, Nr. Canterbury, Kent. Ruined by fire a.m. August 1933.

The derelict tower of Staplehurst smock mill, Kent before and (below) after it was burnt down during celebrations for King George V's Coronation night, June 22 1911.

Billericay post mill was already in a sorry state when it was blown over in a gale on July 27 1929, as the bottom picture shows.

Above: A distant view of the picturesque octagonal smock mill at Egerton, Nr. Ashford, Kent. This picture was taken in 1915 before the peace and tranquility of the area were shattered by a severe gale in December 1919. The upper portion of the mill was blown off, as can be seen in the picture below.

Storm

Many windmills have fallen victim to storms. Designed to catch the wind and therefore usually built in exposed positions, damage to some degree is at sometime inevitable. For example, it is recorded that the Great Storm of 1703, in which the first Eddystone Lighthouse was swept away, 400 windmills were blown down. In addition 800 houses, 7 church steeples and 4,000 trees in the New Forest were also blown down. In many cases damage sustained in a storm has proved too costly to repair and many mills were abandoned and allowed to decay due to this reason. If a mill became tail-winded in a gale it often resulted in disaster. The sails and windshaft assembly rapidly rotating in the opposite to normal direction could be forced out of the neck bearing, with the result that the flailing sails would smash into the mill body and the tail-end of the windshaft tilt up into the cap, or the cap might be blown right off. A tail-winded post mill could well end up blown over because during normal operation these derived some stability from the ladder.

The structural timbers of a post mill or smock mill are large in section and usually of oak, the process of decay therefore may be a lengthy one, a fortunate circumstance which has made possible the restoration of many mills after years of neglect. On the other hand, many another mill has collapsed without warning, sometimes in calm weather conditions, though usually it is a storm which hastens the end of most old mills. The old post mill which stood at Stokenchurch in Bucks. may be cited as an instance in which an apparently sound structure was afflicted by rot in a vital timber. In a booklet *Windmills in the Midlands* published in 1925, J.B. Paddon records a visit he made to the mill and after commenting on 'the giant trestle' and 'blackened oaken beams', goes on to say 'Unless deliberately demolished however, the mill is sure to stand for some years to come'. The mill collapsed the following year in 1926! That this mill, built in 1736, has not survived is especially unfortunate as it was extremely unusual in two respects. Firstly its sails; two of these were cloth covered and quite ordinary, but the other pair which were spring sails, had shutters parallel to the whip, an apparently unique feature. The second point of interest was its trestle, which had three crosstrees instead of the usual pair and six quarterbars, where four usually sufficed. Ironically in view of the mill's eventual collapse the purpose of these extra timbers was increased stability!

The post mills at nearby Bledlow Ridge and Chinnor had trestles of similar design, a fact which would suggest that the same millwright was responsible for all three mills, the feature being so unusual—only two other instances recorded, one in Essex and the other in Notts. The Bledlow Ridge mill was demolished in 1933, but Chinnor mill which was in a very dilapidated condition was dismantled in 1967. Some of its components, notably the bolter, were used in the restored Pitstone mill, but the structural timbers and machinery are stored in a barn in Norfolk and might well one day be re-built in another mill.

Among several mills which have collapsed or had to be demolished because of their dangerous state might be mentioned the post mills at Ash in Kent and Moreton in Essex. Both of these mills were the subject of renovation proposals which became so protracted that eventually it was too late. In the case of Ash mill, a preservation order was made in 1952, but negotiations to repair the mill were still going on three years later in 1955, when the mill collapsed in a storm on October 21st. Moreton post mill, noteworthy for its unusual trestle, which was similar in design to those at Stokenchurch, Chinnor and Bledlow, became so rickety due to neglect that it had to be dismantled, an unhappy end for a mill which 'belonged to the nation'.

Lightning too took its toll. In Dale's *History of Harwich* (1732) we are told 'A violent storm of thunder, lightning and rain on Wednesday 26th June 1728, the Dint of which fell upon the windmill near the town, where it split the main post on which the mill stands, much shattered the weatherboards before, broke one of the sails and much damaged the main part of the mill. By good Providence there was no person in the mill at the time...... The Miller's dog which lay asleep under the mill was killed, but no outward mark of harm appeared upon him'.

In 1829 Toot Hill windmill in Essex was struck by lightning and badly damaged. The sack chain was welded solid and the miller had to have a leg amputated.

Attack

Windmills have often come under attack—during the seige of Colchester in 1648 all the windmills were destroyed by Parliamentarian troops. The Royalists responded by commandeering some millstones and using them in horse mills in the town.

In 1757 a windmill at Clayton, Manchester was destroyed by a rioting mob and in 1795 another at Oldham, also in Lancashire, was extensively damaged in a similar incident.

In Nottingham, Sharpes mill was extensively damaged in the rioting in 1831 and only the arrival of troops prevented the mill from being burnt down.

Removals

Many windmills have been moved from their original site to a new one and for a variety of reasons. The original site may have become unsuitable due to the encroachment of high buildings. The mill might be acquired by a new owner who wanted it on his own land or, in at least one case, at Bromley in Kent, it spoilt someone's view!

The most recent windmill removal and re-erection involved the derelict post mill at Danzey Green in Warwicks. which has been re-sited at the Avoncroft Museum of Ancient Buildings, and is now in splendid structural condition albeit without machinery at present.

Another fairly recent windmill re-siting project which did not materialise as originally intended was that undertaken with the old smock mill which stood at Jolesfield in Sussex. The mill was bought for £250 in 1959 by Mr. Nevver Hickmet, whose intention was to restore it to working order and add it to the collection of old buildings which he had assembled at Gatwick Manor. Unfortunately however Gatwick Manor was acquired by a large catering organisation before this praiseworthy scheme was completed, and the new owners have built a sham windmill of appalling ugliness using nothing from the Jolesfield mill. Regrettably the machinery from the mill lies in the open gradually being eroded by the elements.

Re-Development

The urbanisation of large areas of countryside around towns has resulted in the disappearance of much agricultural land and many windmills. Old maps of what is now the Greater London area record numerous mills, but only two have survived; one a hollow post mill on Wimbledon Common; the other a tower mill, curiously enough in the heavily populated Borough of Brixton. The public gardens in which the Brixton mill stands provide a welcome haven in an otherwise drab neighbourhood.

Threats in the form of planning applications to local authorities to demolish a mill and re-develop its site occasionally occur and often provoke a storm of protest. A mill which perhaps had been neglected for decades is suddenly threatened and local feeling is aroused to the fact that the mill is a visual amenity and the character of the locality will be the poorer without it.

An instance of local feeling, erupting into violence, occurred at the village of Rodmersham in Kent in 1968. Planning permission had been obtained to replace the 135 year old tower mill with two detached houses. The first attempt to demolish the structure was thwarted when the irate villagers surrounded the mill and refused to move. A second attempt by the demolition contractors was more successful. Expecting trouble they arrived at the site in the early hours of the morning, and the first indication the villagers had that the mill was again threatened was the sound of the bulldozers starting up, and several local people barricaded themselves inside the mill. However, the demolition crew, using ladders, quickly forced an entry and in the ensuing mellee one of the villagers was kicked in the head and had a tooth knocked out. Rodmersham Mill was built in 1835 and had worked until 1917 when, like many mills during that period, poor trade forced it to close. The sweeps and machinery were dismantled and the mill became just another derelict.

A similar threat to the spectacular mill at Friston in Suffolk has temporarily at least been averted by a preservation order being placed on the mill. The ultimate fate of this fine mill is however still undecided, but one proposal under consideration is that it be re-erected at the Abbots Hall Museum of Rural Life at Stowmarket.

The tower mill at Rodmersham Green in 1930.

The post mill from Danzey Green during recent re-erection at Avoncroft Museum of Ancient Buildings. The photograph shows the body framework being lowered onto the post.

The windshaft has been installed and the roof ribs are being fitted. Note the absence of triangulation in the body framework, a bad feature which caused the premature failure of many post mills. (Photos: Bromsgrove Messenger).

Above: Mount Ephraim Mill, Ash, 1926. This post mill collapsed in a gale in 1955, when negotiations for its restoration were taking place.

Above: Bredhurst smock mill at the beginning of the century was already in a sorry condition. The ruin became so shaky that it was considered a danger and pulled down. A sad end to a once fine windmill, as our bottom picture shows.

Below: Bredhurst Mill—a sad end to a once fine mill.

Gazetteer

BEDFORDSHIRE

Stevington (992527)

This post mill was acquired by the Bedfordshire County Council in 1951 and restored as part of the Council's contribution to the Festival of Britain. The body has a distinctive appearance with its mansard shaped roof of red painted corrugated iron and prominent flared skirt at the front and sides. The single storey roundhouse is stone built with a steeply pitched tiled roof. Inscribed on a crosstree is the date 1770, probably the year it was built, as maps previous to this date do not record a mill on this site. A second inscription records that the mill was 'rebuilt 1921 P. Keech'. The new timber used on that occasion was generously supplied by Lord Ampthill from his estate. The standard of the work carried out by Mr. Keech, a local builder, was such that only one beam was found to be defective when surveyed in July 1951. The present common sails were fitted in 1958 by Edwin Hole of Burgess Hill. The single pair of stones came from Milton Ernest mill and are mounted in the breast, driven direct from the wooden brake wheel via an iron stonenut. The wooden windshaft carries no tail-wheel, there apparently having never been a second pair of stones in the mill. The tailpole is fitted with yokes against which the miller pushed with his shoulders when turning the mill. The present tentering gear is unusual in being completely in iron, the more usual massive oaken bridge-tree having been replaced by a relatively small section iron component.

This attractive mill, which incidently is the only surviving post mill in Beds., is prominently situated in a cornfield on the Bedford side of Stevington.

BUCKINGHAMSHIRE

Brill (652142)

The pristine appearance of the weatherboarding and brickwork of this attractive post mill disguise the fact that this is England's second oldest dated mill. The date 1668 is inscribed on the post and is said to be the year in which the mill was built.

Old photos of this mill show it with an open trestle and a cart-wheel at the end of the tailpole. During a long period of neglect, most of the tailpole disappeared along with the ladder and the mill deteriorated into a very bad state. Despite its poor condition however the mill has recently been the subject of extensive repair work. This included the fixing of additional supports to the body in the shape of vertical steel props under each corner post, to relieve the rather shaky trestle of the mill's great weight. All the machinery is still in place, including the two pairs of stones. These are underdrift and mounted side by side on a hurst framework in the breast of the mill, having at some time been converted from the original head and tail layout.

Ibstone (769915)

The prospect which this smock mill commands is delightful, standing high on a ridge above the charming village of Turville.

The white painted timber body has a black tarred brick base and is unusual in being twelve sided. The dome shaped cap has a deep petticoat and was originally winded by a fantail, which has long since disappeared, as have the sails. The mill was used as a residence before the last war and some of the machinery was consequently removed. Remaining however are the wind and upright shafts, brake wheel and wallower, great spur wheel and one pair of stones.

The mill was built in the early 19th century and last worked by wind in 1910. Some old maps indicate that a post mill stood on the site previous to this mill.

In 1967 four sails and a fantail were fitted for sequences in the film 'Chitty Chitty Bang Bang', the sails being turned by an electric motor! Unfortunately the sails were removed after filming as it was considered the body was not sound enough to bear their weight.

Turville is in a valley about three miles south of Stokenchurch. The mill is reached by a footpath leading off the road between two cottages opposite the 'Bull and Butcher', then up the steep grassy slope.

Lacey Green (819009)

Standing beside a farm track in the heart of the beautiful Chilterns, this handsomely restored veteran is reputed to be England's oldest smock mill. It is said to have originally been built at Chesham round about 1650 and moved to its present site in 1821. Though many *post mills* were resited, they were simply lifted up bodily and transported elsewhere. The removal of a smock appears to have been less common, perhaps because the structure had to be dismantled for transport and re-erected at the new site. This mill does have roman numerals cut in its structural timbers, a practice commonly used at the time when dismantling and re-erecting wooden structures.

Until recently the mill was in a very decrepit state but, encouraged no doubt by the remarkable achievement with the post mill at nearby Pitstone Green, the Chiltern Society set themselves the daunting task of completely refurbishing the mill; a labour of love begun in the autumn of 1972 when the cap and windshaft were lifted off. The completely restored mill is now open to visitors.

The mill has an octagonal body with four floors, a waggon roof-shape cap with fantail and four sails. Inside most of the machinery has survived. All the gearing is of wood—the brake wheel, wallower and great spur wheel being rare examples of the primitive compass arm form of construction and, perhaps surprisingly in view of the pervading air of ruin which hung over the mill for so many years, all in remarkably good order. Two pairs of over-drift stones are still in place on the first floor, which is just above ground level; the meal floor below being in the brick built base.

In the base one of the timbers supporting the stones carries a date 1785 and some illegible initials. Other inscriptions particularly noteworthy include two of primitive post mills, one incised in a strut on the stone floor and the other in a cant post on the top floor.

This very attractive mill stands in a field behind 'The Whip' public house at Lacey Green, three miles south of Princes Risborough.

Pitstone Green (945157)

This handsome post mill is owned by the National Trust, who were presented with it in 1937 by the former owner Mr. L.J. Hawkins. It has the distinction of being England's oldest dated windmill—the date 1627 is inscribed on a timber of the body framework. The working life of the mill ended abruptly in 1902, when in a severe storm it was tail-winded with disastrous results. Harried by the weather and the attention of vandals, the following 60 years of inactivity left the mill a bedraggled derelict. In 1963 local anxiety for the mill's future resulted in the formation of the Pitstone Windmill Restoration Committee and an extensive re-building programme was worked out. In August 1970, after six years of spare time effort which included the laying of a 250 yard long road to provide access for equipment and heavy materials, new sails were once again turning by wind power.

All the work has been carried out by unpaid volunteers and much of the material used in the re-build was either given free or supplied at a generous discount. All the new timber, oak and larch, was donated by the National Trust and came from the Ashridge Estate, owners of the mill in the late 19th century. The common sails which now grace the mill were made by carpentry students at the Aylesbury College of Education and are authentic in every detail. Its external appearance has been faithfully restored in accordance with an old photograph taken shortly after the mill ceased work. It has a creosoted body and a single storey brick roundhouse. This is a later addition, the trestle having originally been exposed as its weather-beaten timbers clearly indicate. The brick piers on which the trestle is supported show evidence of the mill having been raised at least once; the roundhouse may well have been added when this was done. Additional support in the shape of extra quarterbars and vertical props under the body are also provided to relieve the trestle of some of the great weight. This was done in 1957 to prevent the structure collapsing—it undoubtedly saved the mill but does mean that the sails cannot be brought round to face the wind. Therefore to run the mill, the wind needs to be more or less from the right quarter.

Up inside the body all the machinery is in place. Both brake wheel and windshaft are of wood; the rest of the drive—wallower, upright shaft, great spur wheel and stonenuts is cast-iron and have at some time replaced the former wooden components. No tail-wheel is fitted, both pairs of stones being mounted in the breast and underdrift. Across the rear of the stone floor is a wire machine. This is not the original, which disappeared long ago, but came from the derelict post mill at Chinnor. The tentering gear of both pairs of stones is controlled by a single governor driven by belt and pulley from the upright shaft. The wooden sack hoist bollard is in the very apex of the roof and driven by chain from a crude pulley on the windshaft.

Of particular interest in this mill is the unusually fine carving on some of the original timbers. The main post for example has been beautifully turned to produce ornate mouldings both at the junction of the crosstrees and where it protrudes into the meal floor.

All in all this is a fascinating windmill. It stands prominently in a cornfield south of the village of Pitstone Green, within sight of Ivinghoe Beacon and not very far from Whipsnade Zoo.

Brill mill undergoing repair work.

Quainton (746203)

Towering above the ancient market cross this six floored mill is the tallest in Bucks. Erected around 1830 it is said to have been built in the manner of a chimney stack, the courses of bricks having been laid from inside. It is in a bad state—both fantail and sails have gone, and the cap, which was dome shaped, does not look as if it will last much longer.

Wendover (867083)

This massive octagonal tower was completed in 1804, eight years after construction began in 1796.

The mill ceased working by wind in 1904 because of damage sustained in a storm. Grinding continued however until 1926, the stones being driven by an auxiliary steam engine. The mill has since been stripped of machinery and converted into a house. The cap, said to be the biggest in the country, has been re-built and covered with aluminium. The gallery around the cap, being some seventy feet above the ground, must afford a fine view of the surrounding village.

An unusual feature for this part of the country is the massive four armed cross still in place on the windshaft.

The mill is prominently situated by the A.143.

CAMBRIDGESHIRE

Bourn (312580)

This post mill is accepted as the oldest surviving windmill in the country—an old deed dated 1667 found in the mill about forty years ago referred to its sale in 1636. *(Cambridge Times* 10th June 1932). It is owned by the Cambridge Preservation Society, who were presented with it in 1932 by Mr. and Mrs. Alfred Bossom and Mr. Mansfield Forbes, and it is maintained in good order.

The trestle, as befits a primitive mill, is exposed and carried on short brick piers which are a later addition. A crosstree bears an inscription 'I.U.P. 1874', presumably the date on which the mill was raised onto the piers. The body which is tarred black has a simple straight pitched roof and carries the framework of two common and two spring sails, the latter without shutters. Within the mill most of the machinery is of early 19th century origin, the single pair of stones being directly driven from the wooden brake wheel. The wooden windshaft carries an iron tailwheel which at one time drove a second pair of stones, but these have since been removed. On the spout floor a number of early dated inscriptions can be seen on the timbers of the body framework.

The mill stands behind a small thatched cottage in Bourn Road about half a mile from the village, which is in itself extremely picturesque.

Stevens Mill, Burwell (580665)

This interesting old tower mill is the sole survivor of four windmills which once worked in the village. It is said to be about two hundred years old and was working by wind as recently as 1955. The tower, which has four floors, is built of clunch (a hard chalk), rendered and tarred. There is a handsome ogee cap, four patent sails and all the machinery remains intact inside.

Over (381689)

An attractive small tower mill adapted by its enthusiastic owner Mr. G. Wilson to drive a drill and a saw by wind power. It has two patent sails, a fantail and a beehive shaped cap. Mr. Wilson spent several years restoring the mill to its present fine condition and on days of favourable wind it may often be seen working.

Swaffam Prior (572642)

Two mills—one the shell of an octagonal smock in the last stages of decay, the other a brick tower with four patent sails, a fan stage, but no fan, and a beehive shaped cap. Rather decrepit looking but attractive and structually quite sound.

Great Chishill (413388)

This nicely restored open trestle post mill was built in 1819 using many of the timbers from an earlier post mill which it replaced. It has four patent sails, a fantail on the ladder, and the body has a prominent extension at the rear which houses the bolter. The mill last worked by wind in 1951 and was renovated by Thompsons of Alford, Lincs. in 1966.

Situated beside the B.1039.

Cattells Mill, Willingham (404697)

This very fine smock mill was built in 1828 and worked by wind up until about ten years ago. It is very distinctive in appearance due to its large brick base and very pronounced batten to the weatherboard clad upper part of the body. It has two patent sails, a fantail and an ogee shaped cap, around which is a gallery, an unusual feature. All its machinery is still in place and in good order.

Wicken Fen (563706)

This diminutive drainage mill is reminiscent of the mill at St. Olaves, mainly because of its size and four sided shape. It differs however in having four common sails and a fantail. Built at the beginning of the century, it originally stood at Adventurers Fen in Huntingdonshire. It was later brought to Wicken Fen and stands in 320 acres of natural fenland, which is preserved by the National Trust as a nature reserve.

The pumping mill at Wicken Fen, the last of hundreds which stood in the Fenlands. The sails and tailpole are new. (Photo: A.F. Kersting).

Madingley (408595)

This post mill was brought here from Ellington in Huntingdonshire to replace a mill which collapsed in 1909. The re-erection was completed in 1953, but the mill has never worked commercially on this site. It is of particular interest however as it is a type of mill which was once widely used in the midlands. Unlike the more usual post mill in which the roundhouse is not structurally necessary, the roundhouse of a midlands' type of post mill has a curb on top of the wall on which rollers run. The rollers are carried on a sub-frame fitted on the underside of the mill body, thus the main post and trestle do not bear the entire weight of the mill. The roof protecting the roundhouse is part of the mill body proper and rotates with it.

Inside this mill two pairs of underdrift stones are mounted side by side on a hurst frame in the breast, a practice fairly common in the midlands but not elsewhere. A wooden windshaft carries a composite brake wheel and the wallower and upright shaft are made of cast-iron. An interesting feature peculiar to this mill is the unusual double row of staggered teeth on the brake wheel and wallower; no other surviving windmill has this feature. Two common and two spring sails are intact on the mill and the end of the tailpole is supported by a small iron wheel.

The mill is well cared for and, standing adjacent to a row of stables, gives useful service as a hay store.

This fine mill is situated beside the busy A.45, a couple of miles east of Cambridge.

Six Mile Bottom (589582)

This interesting old post mill was brought to this site in 1846 when the Cambridge to Newcastle railway line was built. There are two dated inscriptions inside; one up in the body reads 1764; and the other 1829 is cut into a crosstree. The earlier date could well be the year of erection as the mill is typical of 18th century practice. Although rather battered in appearance the mill appears structurally sound and has all its machinery intact. Two pairs of overdrift stones are carried head and tail fashion and a flour dresser is situated in the tail of the mill. The brake and tail wheels are wood, but the windshaft is cast-iron and probably replaced the original wooden one when the patent sails were fitted. There is a single storey roundhouse and a tailpole, but the ladder and sails have all but disappeared. Despite it rather shabby exterior this mill is undeniably attractive and evokes an atmosphere of antiquity.

It is prominently situated in a small clearing up a farm track beside the busy A.11.

DERBYSHIRE

Dale Abbey (437398)

This is a midlands' type of post mill preserved in splendid condition, thanks to the owners—the Stanton Ironworks. Its date of erection is obscure, but it is known to have stood here in 1788, at which time it had an open trestle. The stone roundhouse, which carries a curb, having been added in 1844. The mill has four spring sails and winding was by a tailpole.

Machinery includes two pairs of overdrift stones mounted side by side on a hurst frame in the breast, and a wire machine is situated in the very prominent extension at the rear of the body. The windshaft is iron, doubtless a replacement for an earlier wooden one. The wallower is also iron, but the upright shaft and clasp arm brake wheel are of wood.

Heage (366358)

Splendid six sailed tower mill with fine ogee cap. Restored by the Derbyshire County Council in 1967.

Facing page: The curiously named Cat and Fiddle mill at Dale Abbey, Derbyshire. (Photo: A.F. Kersting).

ESSEX

Ashdon (595425)

Post mill two miles from Bartlow. Probably this was originally an open trestle mill which was raised onto its roundhouse about 1830–40. Retains its machinery, including two pairs of overdrift stones, arranged head and tail. Winding was by tailpole. Last worked about 1910. Now privately owned and preserved.

Aythorpe Roding (590152)

Until recently this mill had been in a sorry state, but fortunately Mr. Vincent Pargeter, acting on behalf of the Essex County Council, put in some timely repair work and thus ensured the mill's future for some years to come.

It is a post mill with a single storey brick roundhouse. It has four patent sails, the double shuttered type, but with the shutters removed, and a large fantail mounted on the ladder. The fantail was a later addition to the mill, as proved by the cut off stump of the original tailpole. All the machinery is intact and includes two pairs of underdrift stones in the breast, a bolter in the tail and a cast-iron windshaft. The gearing is cast-iron, with the exception of the brake wheel, which is of composite construction—cast-iron spokes with wooden cogs and rim.

This interesting mill stands beside the main road about five miles from Dunmow. There is also a model of the mill in Colchester museum.

Bocking (763260)

This beautifully restored post mill has the distinction of being the first in the country to be acquired by a local authority for preservation as an historical monument. It dates from about 1680 and originally stood on a site some hundred yards further down the hill. It was moved to its present position on rollers and raised onto the two storey roundhouse, built to receive it, in 1830.

Two common and two spring sails are fitted and there are two pairs of stones. These were originally driven from above, but in 1864 extra gearing was added to convert the drive to underdrift. The stones were regulated by centrifugal governors which are also still in place.

In 1929 the mill was presented to the Bocking Parish Council by Mr. Edward Tabor, whose family had owned it for many generations.

Finchingfield (687330)

Finchingfield is often acclaimed as one of the prettiest villages in England, a description amply justified by the variety of its charming old houses and the beauty of its 14th century Church. Not the least of its unspoiled charms is the old post mill at Duck End, whose century hued timbers have defied the elements since the day it last ground corn over one hundred years ago.

The earliest of several dates cut on timbers inside the body is 1760, so in all probability the mill was built around that date. Although the mill apparently worked until approximately 1870 it was not modified to bring it up to date with the latest windmill practice, as were so many post mills. It retained its common sails, a single pair of stones and oaken machinery until it finally ceased work. Of particular interest is the windshaft, which has not been fitted with an iron poll end, but retains the original mortises in the shaft for the sail stocks, a rare survival.

It is quite a small mill, but one of exceptional attraction. In fact a mill which so evokes an atmosphere of a more leisurely age, that although it should not be allowed to decay into oblivion, one hopes it will not be restored to a state of alien newness, inappropriate to its surroundings.

Fryerning (639008)

A very well kept post mill with its machinery intact, including two pairs of overdrift stones mounted in the head and tail. The brake and tail-wheels are wood, and the windshaft, probably a later replacement, is cast-iron. The post has a date, 1759, inscribed on it. Local tradition insists that the mill is haunted by an 18th century miller killed by one of its sails.

It stands in a private garden about one and a half miles from Ingatestone.

Mountnessing (631980)

Recent repair work has left this fine old post mill in very good order. The narrow lofty body surmounts an unusual sixteen sided brick roundhouse, which has a steeply pitched roof of sheet-iron. The original roof was thatch, but this was removed in 1917 due to the rats nesting in it! *(English Windmills, Vol.2,* Donald Smith).

There are two pairs of underdrift stones mounted side by side in the breast, the framework of four spring sails and a tail-pole. A cast-iron windshaft carries a large brake wheel which is made of wood, as is the upright shaft and all the other gearing. A date 1807 is inscribed on the massive oak crowntree.

Prominently situated on a mound by the side of the busy A.12, about six miles south of Chelmsford.

Orsett (633813)

Octagonal smock mill with brick base. Boat shaped cap with patent sails and fantail. All machinery wood, except windshaft. Three pairs of overdrift stones. Said to have been built in 1674.

Ramsey (209304)

This very tall post mill was originally built in Suffolk and brought to this site in 1842. It has a three floored roundhouse, fantail on roof of body and patent sails. Machinery within includes three pairs of stones, two in the breast and one in the tail, and two upright shafts. It is now derelict.

South Ockendon (604831)

Smock mill: windmills on this site since 1295. This one probably built around 1829. Machinery includes three pairs of stones, boat shaped cap, patent sails and fantail.

Stanstead Mountfitchet (509248)

This tower mill was built in 1787 and ceased working by wind in 1910. The sixty-five foot high mill is today in excellent fettle, having been completely renovated in recent years. The dome shaped cap carries four patent sails and a fantail. All the machinery is intact and includes three pairs of overdrift stones and a bolter.

Stock (698988)

Tower mill. Patent sails; fantail; boat shaped cap. Machinery includes three pairs of stones.

Three miles from Ingatestone.

Terling (764150)

This octagonal smock mill was the last windmill to work by wind in Essex. Originally built at Cressing around 1770, it was moved to this site in 1830. In its working days, three pairs of stones were driven.

Thaxted (609308)

Built in 1804 the condition of this tower mill has been the cause of concern and an appeal has been launched to repair it. The machinery includes three pairs of stones.

Situated behind the Church.

Upminster (557868)

Built about 1801 this magnificent five storied smock mill has been completely restored. Sixty feet in height to the top of the cap, the body is octagonal in shape and carries a stage at the first floor level. The boat shaped cap has an outside gallery and is fitted with a fantail. The cast-iron windshaft was brought from an old post mill to replace the original shaft, which broke in a storm in 1899.

The framework of double shuttered patent sails is fitted, the shutters of which have been removed for safety and stored in the mill.

Four pairs of French burr stones are mounted on the stone floor and are underdrift. Against the wall is a wire machine and an oat crusher, both with belt drive.

On the meal floor can be seen the great spur wheel with final drive to each stone and a belt driven bolter. Also on this floor are the centrifugal governors.

This fascinating mill is situated in a small field at the end of a track leading off the Hornchurch–Upminster Road.

Below: The windmill at Bembridge, I.O.W., which has been expertly restored and is now open to the public.

White Roding (564131)

This tower mill replaced a post mill blown down in 1877. Patent sails, but fantail missing. At present being repaired.

Situated seven miles from Dunmow.

GLOUCESTERSHIRE

Falfield

Derelict tower. No machinery.

Framton Cotterell

Derelict tower. No machinery.

Warmley

This stone tower is part of the remains of the extensive brass works built by William Champion in 1746. It was a windmill used for ore crushing. Nothing but the shell survives. It stands beside remains of a dam and the dried up bed of the mill pond, which once powered the brass works water wheels.

Warmley is on the A.420 just outside Bristol.

HAMPSHIRE

Bembridge, Isle of Wight (640875)

Bembridge mill, the island's sole surviving windmill, stands in an area which, until the reclamation of land from Brading Haven some eighty years ago, was known as Binbridge Isle.

Built around 1700 of local stone it last worked by wind in 1913, but like many other abandoned windmills, the thirty-six foot high tower was put to good use in World War Two as an army observation post. After the war the condition of the mill rapidly deteriorated, and in 1957 the owner, Mrs. Edward Smith, offered it to the National Trust. With the aid of money raised by public subscription the mill has been effectively restored to its present fine condition.

As in its working days, four common sails are fitted, but the distinctive gable shaped cap is no longer thatched as it was originally. During the many years that the mill was virtually derelict, both pairs of stones were removed. The replacements have come from the old tide mill at Wooten Bridge, along with the gantry which was used to lift the runner stones when dressing was required. All the gearing is of wood as is the upright shaft, but the windshaft is cast iron and has at some time undoubtedly replaced an earlier wooden one. The sails were luffed into wind by means of the chain operated winding gear, the eight foot long worm wheel being of particular interest, having been carved from the trunk of a single apple tree.

During the period the mill was undergoing restoration, two dated inscriptions were found. One on the ladder reads 'E. Beker 1746 A.C.' and the other on a timber of the tentering gear is less distinct but appears to be 1701.

This mill is well worth a visit; it is one of England's oldest surviving tower mills and its early wooden machinery is of great interest.

Situated beside the main road between Sandown and Bembridge.

Chalton (716162)

The last of a succession of several windmills which have stood here since the 13th century, this one, a tower mill, is now a ruin. The cap and sails are missing, but inside the brick tower, the machinery is more or less intact. Local tradition insists that the mill is haunted. True or not, viewed at dusk the old ruin has a very eerie atmosphere and the sudden fluttering swoop of a bat from its interior does not encourage a visitor to linger!

Situated prominently on a hill beside the main Portsmouth road, three miles south of Petersfield, an area noted for its outstanding beauty.

HERTFORDSHIRE

Arkley (218953)

A tower mill in good repair. Dome cap with ball finial and gallery. Fantail and framework of four sails. Stands in a private garden.

Cromer (305286)

The date of building this post mill is obscure but windmills have stood here since the 14th century. The present mill has recently been restored to fine condition. It has an octagonal single storey brick roundhouse; a graceful ogee roof to the body; four patent sails and a fantail. Two pairs of stones are mounted in the breast, the mill having been converted from head and tail layout sometime during its long working life. The windshaft is cast iron. The mill ceased work around 1930.

Kings Warden (145233)

Derelict tower, about four miles east of Luton.

Little Hadham (438229)

Octagonal smock in bad state of repair. Built in 1777 and last worked in 1929. Two storey brick base with dome cap. Sails and fantail missing. Upright shaft said to have come from a sailing ship at Yarmouth.

Little Hadham is four miles west of Bishops Stortford.

Tring (915118)

Goldfield mill built *c.*1852. Brick tower with dome cap. No sails or fantail. Ceased work about 1908.

Stands in a private garden beside the Upper Icknield Way.

Weston

Lannock mill. Shell of tower. Built *c.*1860. Ceased work approximately 1922.

KENT

Benenden (821324)

This forlorn old derelict was a fine octagonal smock of the usual Kentish type, but has over the years been denuded of all its internal machinery, the sails and the fantail. It has a two storey brick base, which is quite sound, but the staging and upper part of the body is very decayed and probably will not last many more years.

It is situated in the garden of a private house on the Cranbrook Road, midway between Cranbrook and Rolvenden.

Chillenden (269543)

This mill, although seemingly of primitive type, was in fact one of the last to be built in Kent, the actual year of erection being 1868. The reason for building an open trestle post mill at so late a date is obscure, but was probably one of economy, as some timbers from an earlier mill which this one replaced were used in its construction. The mill today is in immaculate condition, being in the care of the Kent County Council.

It carries four double shuttered spring sails; a tailpole supported by a cart-wheel, and the entire structure is attractively painted white. Unfortunately, inside the machinery is not complete, the runner stones having at sometime been removed, along with the stonevats and hoppers; the gearing and tentering tackle is however intact. The bedstones are mounted side by side in the breast and the drive is from below. All the gearing is in cast-iron with the exception of the brake wheel which is a composite one, wooden cogs mortised into a wooden rim and mounted on a cast-iron spoked hub. Both bridgetrees are also of iron and each is coupled to a separate governor. It is a pity that the machinery of this mill is incomplete but, nevertheless, it is well worth examining for it provides an interesting comparison with earlier post mills, such as Outwood.

The mill stands in a cornfield in the heart of delightful countryside, a few miles east of Canterbury.

Chislet (224679)

Standing with conspicuous ugliness in a rather desolate landscape, this black corrugated iron clad tower is the remains of an octagonal smock mill. According to W. Coles Finch *(Watermills and Windmills,* 1933), it was active in 1905 when he visited the mill, but in March 1916 it was tail-winded in a storm and the cap, complete with the sails, was blown off. The tower now wears a simple gabled roof and although only used as a store it still retains some of the original wooden machinery inside.

A pathway formed of old millstones leads to the front door of the adjacent mill-house which, in contrast to the mill tower, is extremely attractive. Of unusual interest in the attic is the existence of a wooden walled pit dating from the early 19th century, at which time it was evidently used for cock-fighting. Another reminder of that period are the rough bolts and pegs driven into the inner walls of the large chimney. These were common in larger houses in those days, their purpose being to form a crude ladder to enable small boys to clamber up through the chimney to sweep out the soot.

The tower stands in a private garden a short distance from the A.299 near Reculver.

Benenden smock mill, 1927.

Above: Chillenden post mill.

Opposite: The ugly remains of Chislet smock mill, 1935.

Below: Close-up of the spring sails of Chillenden post mill. The shutter frames have had the canvas covering removed. (Photo: A.E.P. Shillingford).

Union Mill, Cranbrook (779359)

Situated in the heart of the Kentish Weald is the ancient town of Cranbrook, noted as the first centre of the cloth weaving industry in the time of Edward III. Dominating the town is the imposing outline of a windmill, Russell's Union mill, the finest smock mill in England.

Built in 1814 by James Humphrey for Henry Dobell, the mill has an octagonal timber body on a brick base and is fitted with two pairs of stones and the usual dressing machines, all of which can be worked by the wind. Further stones and machines are contained in the brick base, driven by an electric motor, which was fitted in 1952 to replace a gas engine.

The post mill shaped cap is seventy odd feet above the ground and is winded by a fantail which was fitted in 1842. Previous to this date winding was carried out manually.

Originally fitted with common sails, the mill now carries the patent type which are adjusted from the iron stage at third floor level by means of the striking chain hanging from the rear of the cap.

This mill has been well renovated in recent years and is in excellent working order—undoubtedly a mill not to be missed.

Opposite: Union Mill, Cranbrook in need of repair.

Facing page: Union Mill, Cranbrook after restoration work had been completed. Kent County Council now own the mill and run it at intervals to ensure that it is kept in working order. (Photo: A.F. Kersting).

Eastry (304545)

At the time of writing some controversy surrounds this old mill, as a recent application to demolish it had been refused. A millwright acting on behalf of the Society for the Protection of Ancient Buildings has surveyed the mill and found that structurally it is basically sound. It is an octagonal tarred smock without sails or fan. Its date of erection is not known, but it is known to have been on the site in 1770. Opinions in the village are divided between those for restoration; those for demolition; and some who could not care less one way or the other. One viewpoint put was that 'Eastry was not a particularly attractive village, but restoration of the mill, complete with a set of sails, would do much to provide a charm which the place lacks at present'.

Below and opposite: Vincent Pargeter and Phillip Barrett-Leonard, two of England's very small band of professional millwrights, applying the finishing touches to the new cap. (Photos: Kentish Express).

Herne (185665)

Windmills have stood here or nearby since as early as 1511. This one, an octagonal smock, was built in 1781 by John Holman to replace a post mill. In 1858 the then all timber structure was raised in height by the addition of a two storey brick base. This was done because trees close by were shielding the sails from the wind.

Typically Kentish in appearance, the mill, which last worked by wind in 1952 is a large one, with a wagon roof cap; fantail and four patent sails. Three pairs of overdrift stones are carried and all the gearing is of wood. The great spur wheel is of particular interest in being of the early compass arm type—not many examples of this form of construction have survived. The brake wheel, which, judging by the position of notches on the inside of the rim, was once also of compass arm design, has been modified by the insertion of a cast-iron hub and spokes. This was probably done when the present cast-iron windshaft was fitted and enabled the original brake wheel, which was formerly carried on a wooden windshaft, to be retained.

This is an extremely attractive working mill, grinding nowadays being carried out by electricity, but it has become evident in recent years that if the mill is to survive, repairs must be carried out. The Herne Society have in fact initiated these repairs by engaging millwrights Vincent Pargeter and Phillip Barrett-Leonard. To date these two young craftsmen have made and fitted a splendid new cap and completely re-tarred the exterior. Further work will be carried out as funds become available.

The mill is easily found, being prominently situated on the hill overlooking the village.

Keston Mill (415641)

This picturesque old post mill stands in the garden of a private house on the edge of Keston Common.

Built in 1716 the weatherboarded body is said to be the largest post mill body in Kent. The remains of the four patent sails are still in place, as are the two pairs of underdrift stones side by side in the breast, plus a centrifugal governor with unusual wrought iron bob-weights. The mill has a single storey brick built roundhouse with a sheet iron roof, and was last worked in 1910. It was however brought back into service in the First World War as an observation post!

This mill is the sole survivor of three windmills at Keston—a smock mill at Holwood was destroyed by fire in 1885; and a tower mill stood at Commonside.

The mill is easy to find. Take the B.265 from Bromley to Keston Common—the mill is on the right, just before reaching the village. Just across the road (on the common) is the source of the tiny river Ravensbourne, which winds its way through south-east London to join the Thames at Deptford.

Kingston (175499)

This derelict brick tower, without sails or cap, would at first sight appear to be an empty shell of little interest. Closer inspection reveals however that although the lower floors have completely gone, the heavy oaken timbers carrying the great spur wheel are still intact in the top of the tower. Made of cast-iron and still retaining all its wooden cogs, the spur wheel is about five feet in diameter and its position so high up the mill indicates that the stones, now missing, were undoubtedly overdrift. Lower down, and ready to fall any minute, is a rusty iron bridge-tree and steelyard dangling from an oaken stump which juts out from the brickwork.

The date of erection of the mill is uncertain, but examination of old maps indicate that around 1850 is the likeliest period. According to W. Coles Finch *(Watermills and Windmills)* the mill ceased work in 1914.

The mill stands in a field in very quiet countryside, about five miles south of Canterbury.

Drapers Mill, Margate (363701)

This, the sole survivor of the many windmills which once stood on the Isle of Thanet, was until recently in a sorry condition. The importance of the mill as an historic link with Thanet farming however has inspired an ambitious restoration project, the major part of which has in fact been completed. The structural framework, weatherboarding and cap were all repaired or renewed in 1969 and the fantail was fitted the following year.

Left: Keston Mill, Kent, 1928.

Facing page: Draper's Mill, Margate, Kent before restoration work commenced. This mill has now been restored to working order.

RAPERS MILLS

Built around 1850 by Holmans of Canterbury, the mill at one time shared the site with two other mills—Little Drapers Mill and The Pumper—both of which have long since decayed and fallen. The existing mill was working by the wind until 1916 and subsequently by gas engine until the advent of the Second World War, during which it finally ceased work.

Meopham Green (639652)

This fine smock mill stands in the picturesque village of Meopham Green, overlooking its famous cricket pitch, and is in excellent condition. The mill's present splendour is due to the extensive restoration work carried out by the Burgess Hill millwrights, E. Hole and Son, on the instructions of the Kent County Council.

The mill, which was built in 1801 from the timbers of a dismantled warship in Chatham dockyard, has five floors and is hexagonal in shape. The cap is the usual Kentish pattern and carries four sails and a fantail. There was a second mill at Meopham known as Priestfield mill, reputedly built in 1769, but no trace of the mill remains however.

The builders of the existing mill were three brothers named Killick, one of whom worked the mill of that name at Strood. The brothers, although not young men, must have been very fit for they walked the eight miles from Strood daily whilst building the mill. *(Watermills and Windmills,* 1933, W. Coles Finch).

Two of the former stones from this mill now rest in the forecourt of the Bank of England, placed there by Sir Herbert Baker, the architect of the building, who lived at nearby Cobham.

To see this attractive mill, take the B.227 road from Wrotham to Gravesend; Meopham Green is about five miles from Wrotham.

The picturesque village of Meopham Green, Kent at the beginning of the century.

Northbourne (331521)

Built in 1848, this tarred octagonal smock mill with a brick base is in the last stages of decay. The cap and sails have gone, leaving the windshaft prominently exposed to the weather.

It is situated a couple of hundred yards to the west of the village.

Rolvenden (837316)

Windmills have stood on this mound for several centuries and the splendid outward appearance of this one hides the fact that it might well be the oldest in Kent. It is a black post mill with a single storey round-house. It has four white painted common sails; a tailpole, but the rear steps have been removed.

Restored in 1956 by local residents, Mr. and Mrs. Harold Barham in memory of their son John, the mill is in good shape and all its machinery is intact. Two pairs of stones are mounted head and tail fashion. Of particular interest is the wooden wind-shaft which is obviously very old—it is square in section and has mortises in which a compass arm wheel once fitted. The existing brake and tail-wheels are of clasp arm construction. It is said to have last worked in 1883.

It is situated beside the Cranbrook Road amidst delightful countryside.

Sandwich (323587)

A map by Robert Morden in William Camden's *Britania* dated 1695 shows a solitary windmill at Sandwich. The site of that particular mill cannot be pin-pointed today, but we may guess that it was a post mill. Later maps of various dates show other mills around the town and W. Coles Finch estimated that six other mills have stood at 'one time or another' in the area.

The sole survivor of these mills is an octagonal smock with one pair of sails and a fantail. The date of erection is not known, but is thought to be late 18th century. Considerable restoration work has been carried out in recent years by Vincent G. Pargeter, assisted by J. Styles. One of the cant posts has been replaced and the remainder strengthened, and the single pair of sails were fitted in 1967. The mill has been recovered with new weather-boarding and further repairs are now being carried out.

Last worked by wind in 1926 the mill, which has its machinery intact, is well worth a visit and is situated beside the A.257 on the west side of the town.

Sarre (260651)

Built in 1820, this octagonal smock mill has been without sails or fan since the First World War. Grinding continued however until about 1930, the stones being powered by a gas engine. Although idle since that date the mill, which like the majority of Kentish smocks is tarred, is apparently in a reasonably good state of repair. The cap, weatherboarding and two storied brick base all look, externally at least, in good order.

The mill stands beside the A.28 a hundred yards up a farm track.

Sarre Mill, 1935.

St. Margaret's Bay (363435)

This well-known smock mill is rather a curiosity, but although it is similar in appearance to many others which stood in the county, it was in fact built in 1929. Its purpose was not to grind corn but to generate electricity. Recent repairs by millwright Vincent Pargeter have restored the mill to fine working order and the sweeps circle briskly in a moderate breeze.

Stanford (129379)

This is a brick tower mill, built in 1857, and it carried four pairs of stones. All that remains today is the tower shell, roofed over with an ugly corrugated asbestos 'cap'.

Stelling Minnis (146465)

Having recently received the attention of millwrights Vincent Pargeter and Phillip Barrett-Leonard, this mill, with the exception of the mouldering outside stage, is in good order. It is a tarred octagonal smock, said to have been built in 1866. Recent repairs have included new double shuttered patent sails and a fantail, all in working order. Adjacent to the mill is a small brick building which still contains the exception of the mouldering outside stage, stones after the mill ceased grinding by wind around 1925. The driving belt from the oil engine is carried up into the mill body via a weatherboarded casing connecting the mill with the engine house.

This mill is situated in a small field south of the village.

The smock mill at St. Margaret's Bay.

Opposite: Stelling Minnis Mill in a sorry state before being restored to the fine condition illustrated below.

The windmill at Willesborough. (Photo: A.F. Kersting).

West Kingsdown Mill (582622)

This well renovated smock mill stood originally at Farningham and was moved to this site in 1880 to work alongside the post mill which stood there.

The post mill is reputed to have been brought from Sevenoaks and to have had a 17th century date carved on the post. Unfortunately it was completely burned down in 1909.

The smock mill has an octagonal body on a brick base and carries the framework of common sails. The cap is of typical Kentish shape with a seven bladed fantail. The mill was worked until 1929, after which date it gradually deteriorated into a bad state of repair. A preservation order was put on the mill in 1952 and subsequently it has been restored to its present fine condition. The mill is maintained by the Dartford Rural Council.

It is easy to find as it is prominently situated alongside the A.20, halfway between Farningham and Wrotham.

Borstal Hill, Whitstable (105652)

Said to have been erected in 1815, this black octagonal smock mill had a house built on to it in 1918 and all the internal machinery was removed. Nowadays it forms part of a hotel and is a conspicuous landmark above the town. Although known in its later working days as the Black Mill, it was originally painted white, but when renovated in 1885 the new weatherboarding, with which the mill was recovered, was tarred black, a more durable finish for a situation so near the sea. Trinity House had to be advised of the change in the mill's appearance as it was used as a navigational landmark by seamen. (*Watermills and Windmills,* 1933, William Coles Finch).

Willesborough (032422)

In danger of falling into a bad state of repair, this large octagonal white smock mill has recently begun a new lease of life. It has been acquired by Mr. T. Robbins, a windmill lover, who intends to use part of the interior to carry on his business of organ building.

Built in 1869, the mill has a two storey brick base; the stocks of four sails and the remnants of a fantail. In its heyday it carried five pairs of stones. Although the mill looks rather decrepit, due mainly for want of a coat of paint, it is in fact structurally quite sound, most of the original machinery being intact and in good condition.

Situated at the end of Mill Lane, near the junction of the A.292 and the A.20.

Stocks Mill, Wittersham (913273)

This attractive old post mill is very well kept and, pristine in white paint, enhances an already delightful garden. The earliest of several dated inscriptions on the main post reads 1781 but whether or not this records the year of erection is not known.

A splendid set of spring sails complete with shutters grace the mill, but no machinery remains inside except for the windshaft, brake and tail-wheels. The windshaft is cast-iron and square in section and has probably replaced an earlier wooden one. Both the brake and tail-wheels are of wooden clasp arm construction with each cog mortised into the rim.

A rare survival on what was once the meal floor is an 18th century firemark of the Royal Exchange Insurance Company affixed to the crowntree. Windmills, being particularly prone to destruction by fire, often carried a firemark, displayed on the outside of the mill, because the insurance company's firemen would not have attempted to quell a blaze if the building on fire did not carry the company's mark.

Another interesting detail in this mill is the sliding trap door on the stone floor, which is designed to close automatically by a rope, pulley and counterbalanced weight action.

Although bereft of machinery, a visit to this mill in its magnificent setting is very enjoyable.

LEICESTERSHIRE

Kibworth Harcourt (689945)

This interesting old mill is the last post mill remaining in the County but, fortunately, local interest ın the mill is strong enough to get a restoration programme under way. It is hoped this will result in the mill being in working order.

The earliest date noted in connection with the mill is 1711 and it was last worked by wind power in 1912. It is a post mill in the Midlands tradition i.e. the brick roundhouse wall has a curb which supports the body via rollers, thus relieving the main post of some of the weight. A single pair of sails is carried and most of the early machinery is intact, including a wooden windshaft with iron poll end, wooden clasp arm brake, tail-wheels and two pairs of stones. The tail-wheel has been converted from a compass arm wheel. The mill was brought into wind with the tailpole.

A picturesque mill in a setting enhanced y the nearby thatched farm-house.

Morcott (931002)

Built in 1875, this nicely restored brick tower mill has a fine ogee cap, four patent sails and a fantail. No machinery remains inside the mill which, in fact, has been converted very skilfully into a house.

Facing page: Maud Foster Mill, Boston, restored in 1953 with money donated by Albert Reckitt, a grandson of the original owner. (Photo: A.F. Kersting).

LINCOLNSHIRE

Alford (457766)

This fine tower mill was built by Oxley of Alford in 1837 and is of exceptional interest in that it is the only surviving five sailed mill working commercially. Its excellent state of repair is due to the enthusiasm of its owner, Mr. F.C.L. Banks, who acquired the mill in 1957, at which time it was in poor condition.

It has six floors, five patent sails and a fantail. The ball finial which tops its very elegant ogee cap is approximately sixty feet above the ground. There is a gallery at third floor level.

This mill is typical of many such mills that once stood in the county, and it may be seen working occasionally.

Maud Foster Mill, Boston (333447)

Slender brick tower mill of six floors with five sails and fine ogee cap. The fan stage survives, but fantail is missing. There is an outside gallery at the third floor level. The mill takes its name from the Maud Foster drain beside which it stands.

Burgh-le-Marsh (504650)

Built in 1844 by the same millwright, Oxley of Alford, who built the similar five sailed mill at Alford, this fine tower mill has been restored to working order by the Lindsey County Council. It is a slender brick tower of five floors, has five clockwise sails, an ogee cap and a fantail. It is a particularly interesting mill to examine, as apart from the machinery being intact, one stone casing has been sectioned and another pair of stones is separated to enable the workings to be more easily appreciated.

Heckington Mill (145436)

This beautifully restored tower mill is noteworthy in having eight sails. Although several eight sailed mills once stood in the county this is the sole survivor. The mill was originally built in 1830, but subsequently burnt out. The shell was acquired by Mr. John Pocklington and rebuilt in 1892, using the eight sails and fantail from the dismantled Tuxfords mill at Boston. The mill worked continuously until 1943, when Mr. Pocklington died. Kesteven County Council acquired the mill in 1953 and they engaged Thompsons of Alford to renovate it. Two of the present sails came from the old Bolingbroke mill and another pair from Wainfleet.

This unique mill stands beside the B.1394 and is a prominent landmark.

Trader Mill, Sibsey (340510)

This fine seventy-five feet high tower mill was built in 1877 to replace an earlier post mill. Four of its six original patent sails remain. It has an ogee cap with ball finial in the Lincolnshire tradition, and a fantail. All the internal machinery, including iron windshaft, gearing and three pairs of overdrift stones, is intact. Particularly attractive is the elegant wrought-iron stage. The sails are carried on a six armed cast-iron cross.

Situated beside the B.1184.

Skidby Mill (021334)

This local landmark is in excellent order as a result of money provided by Beverley Rural District Council and the Department of the Environment. Recent repairs have been made to the sails and some of the surrounding buildings have been converted into an agricultural museum.

Wrawby (026088)

This is the sole surviving post mill in Lincolnshire, a county noted for its fine tower mills. It is in splendid condition, having been renovated in recent years by local enthusiasts with the assistance of Messers Thompson of Alford. The exact date of erection is obscure, but the mill was known to exist in 1800 and was probably built around 1790. The mill has had a long working life with a succession of millers, and only finally ceased work when two sails were lost in 1940.

In the twenty years of inactivity which followed, the mill's condition gradually deteriorated until, in danger of becoming irreparably derelict, a decision was made to save it. In 1962 it was photographed, carefully measured and dismantled. The machinery was renovated and stored to await the replacement of the reconstructed body onto the post. Renovation was finally completed in 1965. Oak for the renovation came from the Brocklesby and Elsham Estates, generously provided by the Earl of Yarborough and Captain Elwes. Financial aid in the shape of Ministry of Works, County Council grants and public subscription was also received.

The mill today is a fine example of a type once common in the Midlands and north-east of England. The four spring sails are carried on a four armed cross, which was the normal practice in Lincolnshire. For most of the mill's working life common sails were carried; these were replaced in 1917 with spring sails which came from the windmill at Laceby. Two pairs of stones are mounted head and tail fashion, the drive is direct with all gearing in wood. The roundhouse, which was a later addition to the mill, houses an interesting collection of milling bygones. A curious feature of the roundhouse is its roof, which instead of being carried by the roundhouse wall and supported at the apex by the quarter-bars, is fixed to the mill body and rotates with it.

This is a fascinating mill which can be seen working occasionally when the wind is favourable, when souvenir bags of stone ground flour may be purchased.

Situated at Wrawby in a pleasant clearing beside the A.15.

The windmill at Heckington. (Photo: A.F. Kersting).

LONDON

Brixton (306743)

This elegant tower mill with its boat shaped cap is a pleasant anomaly amid the rather drab streets of this south-east London borough.

The mill was built in 1816 and worked by wind until 1862. During those years the area around the mill was open fields, but began to be built up with housing about 1850. In 1902 a gas engine was installed to power the stones and the mill was worked until 1934 by this means. In 1964 the mill was restored to its original condition by the Lincolnshire millwrights Thompson and Son, on the instructions of the London County Council, who bought it in 1957. Some of the machinery fitted is from a dismantled mill at Burgh-le-Marsh, Lincolnshire, but the sails were specially made.

The brick built tower is thirty-nine feet high and carries two common and two patent sails. The cap was winded manually by chain via pulley and cog-wheel meshing with a rack fitted around the curb.

Situated in the very pleasant Windmill Gardens in Blenheim Gardens, Brixton Hill, Lambeth, SW2.

Wandsworth Common (267745)

In a metal railed enclosure on Wandsworth Common stands a small hexagonal wooden tower. Probably most of the hurrying passers by would be surprised to learn that this tower is the remains of a windmill. It was built in 1837 to pump water from the nearby railway cutting back into the ornamental lake from which the water was draining. In its working days the mill was fitted with double shuttered patent sails and a fantail. These were removed when the lake was drained and filled in. Today the body is but an empty shell hiding the existence of a well underneath.

Situated at the junction of Windmill Road and Spencers Park Road.

◆ The windmill at Skidby. (Photo: A.F. Kersting).

Wimbledon (230725)

This windmill is strictly speaking a composite mill, a small post mill like cap turning on a curb atop a tapered body, which is built above an hexagonal two storied brick base. The mill was re-built in this form in 1893 by Saundersons of Louth, when it was found during renovations that the original structure was irreparably rotted. The original mill, which was reputedly built around 1817, was a hollow post mill, the drive from the windshaft being taken down through the hollow post to the stones in the body below. The brick base housed the miller and his family and, in addition, provided storage space for the grain.

The mill ceased work in the 1860s and the following thirty years of inactivity left the structure in a poor state of repair. The cost of the re-building in 1893 was met by public subscription, possibly the first windmill to be saved from demolition by this means. This is also another mill with an association with the literary world, Lord Baden-Powell having wrote part of his *Scouting for Boys* during a brief stay here.

With the exception of the windshaft, brake wheel and fantail drive, no machinery remains in the mill. The centre body portion which in the original contained the stones, is now simply a sheet-iron covered framework.

In appearance however, the mill is substantially similar to the original; the sails are shorter and the tapered body is perhaps more slender, but the mill is still an attractive addition to an already delightful common.

The windmill on Wimbledon Common has been expertly restored and is now a delightful meeting place, with its adjoining tea-rooms and park facilities. (Photo: A.F. Kersting).

Below: The old mill, Wimbledon before renovation. During and after the 1914–18 War when standard bread was all the rage, the miller took stones from here to his mill at Kingston.

NORFOLK

Berney Arms (465050)

Built in the latter half of the 19th century this seventy feet high tower mill is one of the many hundreds which were once busy draining the marshes. Typical of Norfolk windmill practice, it has four patent sails, a boat shaped cap and a fantail. Around the body at third floor level is an iron stage. Two features make this mill unusual. First, its scoop wheel is carried in a casing some distance from the mill tower, the drive being via a long shaft (most drainage mills had the scoop wheel against the side of the tower). The second point is that this was a dual purpose mill which at one time carried stones on the second floor which were used to grind cement clinker.

With the white painted sails, cap and fan contrasting strongly against the tarred tower, one of the tallest left in the county, the mill is a prominent landmark for many miles around, but one which can only be reached by rail or boat.

Billingford (167785)

A fine five storied brick tower mill which was working until 1956. It is now owned by the Norfolk County Council and has recently been restored. It carries four patent sails, a fantail and a fine boat shaped cap.

The mill, which was built in 1860, is situated on the common just outside Scole.

Blo Norton (011790)

A derelict brick tower mill. Stocks of two sails. Half collapsed beehive shaped cap. Site appears to be a scrapyard. Tops of numerous oil drums nailed to tips of sail stocks and fan stage timbers!

Burnham Overy (838438)

Built in 1816 to supplement the watermill, which stands at the foot of the hill, this attractive tower mill belongs to the National Trust. The mill worked by wind until 1914, but damage sustained when it was tail-winded in a storm that year ended its career as a corn grinder and it was subsequently stripped of its machinery and converted into a dwelling. Externally however this mill retains its sails, fantail and gallery which pristine in white paint make a brilliant contrast with the tarred brickwork of the tower.

Situated by the side of the A.149, about ten miles east of Hunstanton.

Caston (951982)

Brick tower mill at present undergoing repairs.

Cley-next-the-Sea (045441)

This large tower mill is very attractively sited almost at the water's edge. It has four patent sails, fantail and a gallery. It last worked in 1917. Only the incongruous chimney from the cap proclaims that this is another skilful house conversion.

Denver (605012)

This cement rendered brick tower mill was built in 1835. It is in good order with four patent sails, fantail and a gallery. Ceased working by wind power during the Second World War. Nowadays grinding is carried out by an auxiliary engine.

Garboldisham (002805)

Neglected for decades, this 18th century post mill was literally about to collapse. The roundhouse wall had gaping cracks in several places and the body had over the years been reduced to a broken skeleton. Recently however the mill has been acquired by Mr. Adrian Coleman and Mr. Phillip Unwin who, undeterred by its battered state, plan to restore it to working order and produce wholemeal flour commercially! Millwright Mr. Phillip Barrett-Leonard was engaged to attend to the mill's sub-structure. He has strengthened the brick piers and renewed the foundations, and a completely new trestle has been built. Although the mill has suffered over the years at the hands of vandals, the machinery is fairly intact and includes two pairs of stones in the breast. The windshaft, upright shaft and gearing are all made of iron. Two sails are still in place but, not surprisingly, in bad condition. Although much work remains to be done and it is estimated that the entire project will take four years, the immediate future of this interesting mill is ensured.

The mill stands beside the B.1111, half a mile south of the village and almost on the Suffolk border.

Hickling (408230)

Brick tower mill without sails or fan.

Stubbs Mill, Hickling (437220)

Derelict drainage mill with machinery intact, but all in poor state.

Horsey (458221)

This drainage mill was built in 1912 to replace the earlier mill which was in bad condition, having lost its cap in 1895, and subsequently left to deteriorate. Most of the machinery from the earlier mill was utilized in the successor. It is in good order today, being in the care of the National Trust.

Situated by Horsey Mere on the B.1159 coast road, three miles north of Winterton.

Paston (316358)

Brick tower mill in good order, having been repaired by the Norfolk County Council. It has four patent sails, fantail and boat shaped cap. All machinery is intact.

Stalham (364239)

Hunsett drainage mill on the river Ant is a well maintained brick tower but without machinery.

Stracey Arms (442090)

A mill very similar to the one at Horsey Mere, but in the care of the Norfolk County Council.

Situated on the bank of the river Bure at the point where the river and the Acle New Road run parallel for a short distance, which is about six miles from Great Yarmouth.

Sutton (396239)

This is the tallest mill left standing in the county, unfortunately without sails or fan.

Thurne (400159)

This elegant white drainage mill is probably the best known mill in the county, situated as it is in the heart of the Broads. It is maintained in good order by the Norfolk County Council and is complete with four patent sails and fantail.

Situated beside the river Thurne at Thurne dyke.

NOTTINGHAMSHIRE

North Leverton (776820)

This is one of the few windmills in England which still works by wind power. It is a small three storied tower mill built of brick and tarred. Three pairs of overdrift stones are carried on the first floor, two pairs of which can be driven by wind power. The third pair may be driven by an auxiliary engine when there is no wind. The mill has a full set of patent sails, a fantail and an attractive ogee cap, all painted white.

Although the mill works virtually every day it does not make a profit, all the income from grinding being absorbed by running and maintenance costs. Fortunately the group of shareholders responsible for the mill are happy to keep it running as a useful and attractive asset to the local community.

Situated on the eastern edge of the village within sight of the power station.

OXFORDSHIRE

Bodicote (457367)

Built in 1865 this tiny open trestle post mill has an interesting history. It was designed and erected by Henry Warriner, apparently as a plaything, for it is too small to have worked commercially. Warriner is famed as the inventor of the first screw driven boat, the machinery from which survives in the Science Museum. The trials of this boat, named 'Firefly', were carried out around 1860 on the Sorbrook, which runs at the foot of the hill on which the mill stands. The brook had to be widened for the purpose and an island was formed, which was later used as a test site for agricultural machinery. At that period an island was a wise choice, for it enabled the then revolutionary new machinery and implements to be more easily protected against marauding gangs of protesting farm workers, who were ill disposed towards any supposed threat to their livelihood.

The mill is in complete working order having recently been fully restored by John Lampitt and Company of Banbury, acting on behalf of Mr. Gibbard the tenant of Grove Farm. Inside the small body the direct drive to the single pair of stones is similar to that in a normal size post mill. The whole of the rear of the body opens to allow access to the interior, which has just enough room for one person. Four common sails are fitted and the mill can be luffed into wind by tailpole.

Rather curiously five marble headstones, said to mark the graves of dogs, stand near the mill. *(Banbury Guardian,* lst December, 1966). The inscriptions read, Fritz 1805, Dasil 1839, Dot 1877, Dackle 1892 and Gypsy 1898. No other information regarding dogs or owner has so far been found.

Littleworth Common (589053)

Stone built tower, unusual in being octagonal in plan. Derelict with no sails or fan. During his visit to the mill in 1925, J.B. Paddon noted a date 1764 stamped on the poll end.

Milton Common (637025)

18th century brick built tower mill. Derelict but still has remnants of the common sails and a crumbling fan stage.

Stands in a field beside the B.4013.

SOMERSET

Ashton Mill, Chapel Allerton (414504)

Of the many windmills that once stood in the county, Ashton Mill alone remains complete with sails and machinery. The splendid condition of this mill is due to the generosity of former owner Mr. C.C. Clarke, who had it fully restored before presenting both the mill and field to the Bristol City Museum in 1966. The mill is a circular limestone tower carrying four common sails. Winding is by chain and pulley. Last worked by wind, for grinding grist, in 1927.

Chapel Allerton is about four miles south-west of Cheddar.

Brockley Combe (473661)

Derelict stone tower on the edge of a forest plantation.

High Ham (433305)

Thatched stone tower recently repaired by the National Trust.

Kenn (411697)

Derelict tower.

Locking (358617)

Derelict tower.

Uphill (318583)

Tower converted into observation platform.

SUFFOLK

Bardwell (941737)

Derelict brick tower mill. Beehive shaped cap. Crumbling fan stage. No sails.

Drinkstone (964622)

Very attractive post mill with roundhouse, built in 1685. Despite its great age it is still in working order. Clockwise sails, two cloth covered and two spring. All machinery intact, including two pairs of stones mounted head and tail. The mill stands on a mound behind the mill house, beside which is the empty shell of a smock mill. When this latter mill was built and last worked is not known.

Situated midway between the villages of Drinkstone and Woolpit.

Eye (139744)

Standing like a gibbet in the midst of a clump of wild undergrowth is the post of this once fine mill. It is said to have last worked in 1929, but was then abandoned to the destructive power of the elements. Deterioration must have been unusually rapid because the mill collapsed suddenly in 1953. Penetration of the tangled brambles and creepers forming the outer defences of the ruin is difficult, indeed painful, but once within one finds the red brick roundhouse still intact and a shaky trestle supporting the post. Inscribed on the post is a date 1779, probably the year in which the mill was built. Millstones, gearing and a jumbled assortment of parts lie in disarray where they fell, either in the bottom of the roundhouse or, like the windshaft and brake wheel, half concealed amongst the nettles outside.

This sad but intriguing old ruin is situated on the edge of a field on the north side of the Yaxley Road just outside Eye.

Framsden (192598)

In spite of its impressive height, one comes upon this old post mill quite suddenly, for it stands in undulating countryside amid an abundance of trees. From whichever direction the mill is approached as one gets near it disappears from view, until all at once there it is.

Built in 1760, the mill, like many post mills, began life with an open trestle, four cloth covered sails powered the stones and winding was by tailpole. In 1836 a new owner, John Smith, acquired the mill and had it extensively modernised, the work being carried out by the well-known Suffolk millwrights and iron founders Whitmore and Binyon. The mill was jacked up eighteen feet and the three storied roundhouse built beneath it. The common sails gave way to the more advanced patent sails and the tailpole was replaced by a fantail fitted to the ladder. Within the mill body all the machinery was also brought into line with current practice. At that time cast-iron was becoming increasingly used for mill machinery and the direct gearing of the old post mill was converted to an indirect drive using mainly iron components. The cogging on the brake wheel and stonenuts however was in timber, each cog being separately mortised in. The wallower and great spur wheel are entirely in iron, thus all the meshing cogs are wood to iron which was said to be both smooth running and quiet. The stones were re-sited side by side in the breast, the final drive from the upright shaft being from below. The earlier hand tentering was supplanted by a centrifugal governor which controlled both pairs of stones. The governor is carried on the stone floor, instead of the more usual position below it, the pulley from which the drive is taken being mounted on the upright shaft immediately below the wallower.

Although the mill had stood derelict since the 1930's, when it was last worked, it is now in remarkably good order. The present owner, Mr. Ablett, had considered demolishing the structure but happily was persuaded instead to have it repaired, all the work having been carried out by a small band of volunteers.

Generous support in the form of financial aid came from the East Suffolk County Council, the Suffolk Preservation Society and Lord Tollemache. Most of the work has been completed and is of a very high standard of workmanship, the re-cogging of the brake wheel (which occupied three hundred hours of spare time effort!) being particularly noteworthy. One pair of stones complete with its tentering gear is in working order and capable of grinding corn. At present however the wind must be from the right quarter, as the mill cannot be turned on its massive oaken post until the ladder and fantail have been renovated. Access to the mill body via the present rather ricketty ladder is quite an exciting climb, for with the exception of perhaps Saxtead and Friston mills, both also in East Suffolk, this must be the tallest surviving post mill in the country.

The mill dominates an attractive cluster of old buildings, one of which, a barn, houses Mr. Ablett's intriguing collection of country bygones, things such as agricultural implements, early mouse-traps, even an array of coffin plaques! Aside from the mill, this remarkable assortment of fascinating objects alone would merit a visit.

This haven of rural quietude, Framsden, is situated eleven miles east of Stowmarket on the A.1120.

Friston (412602)

Visible for miles around, this spectacular mill is fifty-one feet high to the apex of the roof, which makes it the tallest post mill left in England. Only two patent sails of the original four remain, as the other pair had to be removed in 1943 because they had become unsafe. The mill has not worked for some years and is beginning to look bedraggled. In fact an application was made to demolish it, but happily this has been thwarted, for the time being at least.

All machinery is still in the mill. There are three pairs of stones, two in the breast and a third pair at the back, driven off the tail-wheel. An iron windshaft carries iron brake and tail-wheels with inserted wooden cogs. There is a flour dresser and a jog-scry, whilst lower down in the exceptionally tall three storied roundhouse is an auxiliary oil engine. The huge fantail is mounted in the approved Suffolk manner on the ladder which, incidentally, is even more unnerving to climb than the one at Framsden Mill!

This is an outstanding windmill and well worth a visit. It stands in the mill yard in the centre of the village of Friston, which is midway between Saxmundham and Aldeburgh.

Herringfleet (465976)

Maintained in good order by the Suffolk County Council, this octagonal smock drainage mill has four common sails and a braced tailpole carrying a small hand winch for grinding. Built around 1850 the mill has only ceased working in comparatively recent years.

Situated beside the river Waveney about one and a half miles south of the St. Olaves drainage mill, with which it makes an interesting comparison. It is difficult however to reach this mill except by boat.

Holton St. Peter (403773)

This nicely restored post mill is attractively sited on a mound above the village. With the exception of the wooden brake wheel and a cast-iron windshaft, there is no machinery remaining inside. It is however well worth a visit, as it provides the rare opportunity (on a windy day) of seeing a post mill luffed into wind by its fantail. When this mill was built is uncertain, but the earliest of many dated inscriptions inside reads...Swan 1749.

Holton is about half a mile from Halesworth, which is on the A.144 eight miles south of Bungay.

Pakenham (931695)

Built in approximately 1810 this, one of the two surviving tower mills in Suffolk, is one of England's few working windmills and its circling sweeps are familiar to millions through the medium of television.

As befits a working mill, it is in excellent fettle with its four patent sails, fantail and dome shaped cap painted white which, contrasting with the black tar on the body, makes an attractive combination.

The mill stands beside the A.143 about four miles north-east of Bury St. Edmunds and in favourable weather conditions visitors may well be rewarded with the sight of the sweeps turning briskly.

The windmill at Pakenham. (Photo: A.F. Kersting)

St. Olaves (Priory Mill) (457998)

Small, square in plan, smock drainage mill built in 1910. Four patent sails and fantail. In a bad state of repair; parts of the sails have fallen and lie rotting in the undergrowth. Large areas of weatherboarding have become detached, exposing the rudimentary machinery to view. However a very interesting and attractive mill.

Situated beside the river Waveney just a few hundred yards from the bridge where the A.143 crosses the river.

Saxstead Green (254645)

Charming Saxstead Green is a perfect setting for this magnificent post mill. Maintained in complete working order by the Ministry of Public Buildings and Works, this white painted mill is typical of the many windmills which once graced the Suffolk countryside.

Built sometime during the 18th century the mill has undergone many modifications during its working life. The body which originally surmounted a low single storey roundhouse was raised onto the present three storied structure in the 1850's, doubtless to make better use of the wind. The fantail winding gear was probably fitted at the same time; previously the mill was winded by tailpole. Patent sails have replaced the earlier common sails and the cast-iron windshaft was formerly of wood. Two pairs of overdrift stones are mounted in the breast of the mill, both driven from the brake wheel via wallower and upright shaft. No tailwheel is fitted.

The mill ceased work during the Second World War and was placed in the guardianship of the Ministry of Public Building and Works in 1951 by Mr. S.C. Sullivan, son-in-law of the last miller, the late Mr. A.S. Aldred. Now preserved under the Ancient Monuments Act, the mill was completely renovated between 1957 and 1960 and is probably the finest in the country. An excellent detailed description of this mill is given by Rex Wailes in his book *The English Windmill.*

Saxstead Green is situated at the junction of the B.1120 and the B.1119, about two miles from Framlingham.

Stanton (971733)

Post mill with two storey roundhouse and covered porch at the rear of the body. The mill is unfortunately in poor condition, but two sails remain and all the machinery, including two pairs of stones in the breast, has survived. The fantail was mounted on the rear of the roof, but has now disappeared, but remains of the staging can still be seen. An interesting mill and one which deserves repairs.

Situated within sight of the A.143, about three and a half miles north of Pakenham windmill.

Syleham (214778)

This post mill worked up until the mid 1950's but is now, unhappily, derelict. It is however substantially intact with two patent sails, a fantail on the ladder and all its internal machinery.

Situated about three miles east of Billingford tower mill. Permission is needed to see this mill at close quarters as it stands at the end of a farm track some distance from the road.

Thorpeness (468599)

A pair of well-known landmarks near the coast at Thorpeness are the five storied 'House in the Clouds' and the adjacent windmill. Although the former has the appearance of a folly it is in fact a water tower supplying the village, the windmill used to provide the means of raising water up into its tanks.

Built at Aldringham in 1803 the mill was brought to its present site in 1924 and converted from its original purpose of corn grinding to water pumping. A typical Suffolk post mill, the white painted body carries four patent sails and a fan on the ladder. The single storey roundhouse (actually square in plan) houses the pumps which are driven by a spindle passing down a hole drilled through the centre of the post. The mill, although no longer used, is kept in good repair.

It is situated a few hundred yards from the delightful Meare, an expanse of water some sixty acres in area.

The windmill at Thorpeness with the 'House in the Clouds' on the left. (Photo: A.F. Kersting).

Walberswick (487737)

An atmosphere of desolation pervades this capless brick shell of a drainage mill. Surviving machinery consists of upright shaft, pit wheel and spur wheel. Still in place also is a braced tailpole with wooden wheel which ran against cemented track on brickwork of tower. Crank and geared winch on end of tailpole was used in conjunction with short posts set around mill. Sail stocks lie half concealed in the massed nettles, along with cast-iron windshaft.

Woodbridge (264493)

Buttrums mill is a well preserved brick tower mill with four patent sails, an elegant ogee cap with ball finial and a fantail. White paint gives the sails, fan and cap a pristine appearance and, in conjunction with the mill's elevated position on the outskirts of the town, help to make it a prominent landmark.

Situated in a clearing behind a private house in Birkett Road.

The mills on Outwood Common in 1927. Alas, the smock mill collapsed in 1962, but the post mill has been restored to working order.

SURREY

Lowfield Heath (270399)

For how long this old post mill has stood here is not known, but Roques map of 1765, the first large scale map of the county, shows a mill symbol at Lovel Heath. It last worked by wind power in the 1880's, and the subsequent years of neglect plus depredation by vandals, had left it a dismembered derelict. Happily however, such was the quality and strength of the structural timbers, the trestle and body framework were sound enough for a restoration project to be put in hand. A great deal of repair work has been completed, all the rubbish and rotten timber has been cleared away, new cap ribs fitted and all the defective weatherboarding renewed. Restoration will continue as funds permit but the fine work carried out so far has ensured the mill's protection from the weather for many years.

Unfortunately not much machinery has survived. There were two pairs of stones, mounted head and tail fashion, of which only the bed stones remain. The brake wheel is a wooden clasp arm type and the windshaft and tail-wheels are cast iron, probably later replacements. The sails, which have been missing for years, were the patent type, but the remains of the striking gear is still in place. There is a single storey brick roundhouse and a tail-pole.

The mill is prominently situated between Gatwick Airport and the A.23 road.

Outwood Mill (328456)

In one of the loveliest parts of Surrey, just a few miles from Reigate, lies the delightful common at Outwood. Here one may be confronted by an extraordinary spectacle—the circling sails of a windmill. This is England's oldest working windmill. It is a post mill, which the original deeds show to have been built in 1665, and tradition has it that the great fire of London could be seen from its roof. The black weather-boarded body surmounts a single storey brick roundhouse which was probably a later addition to protect the formerly exposed trestle. The roundhouse contains an interesting collection of milling bygones and photographs recording the mill's history.

Four spring sails are fitted and drive two pairs of stones, one pair peak and one pair burr; each pair is controlled by a centrifugal governor and mounted on each stone spindle. The mill is turned to face the wind by means of the tailpole and is so finely balanced that one man can quite easily turn it. When working, the area beneath the sails is roped off for safety as the sails come very near the ground. On Sunday afternoons, if the wind is favourable, the old mill is busy grinding and visitors may purchase bags of the stone-ground flour, though the bulk of the mill's produce is taken by health food shops! The mill is owned by two brothers, Mr. G. and R. Thomas, who forsake their more normal weekday occupations to work it on Sundays. Visitors are welcome and allowed to wander at will throughout the mill's interior, so every aspect of the machinery may be observed working. The stone floor is particularly fascinating as one has to clamber about in close proximity to the massive wooden gearing and the mill feels alive with energy as the great weight of the sails cause the structure to sway and vibrate in a slightly alarming manner.

In about 1800 a smock mill was built on this site and was intended eventually to supplant the post mill. It last worked in 1914 and then fell into disrepair until it finally collapsed in 1962, leaving the present mill in undisputed command! Some of the timbers from the smock mill have been used in the construction of an outbuilding near the post mill. A further outbuilding houses a number of early carriages and farm implements, whilst doves, guinea fowl and a donkey are further attractions particularly popular with the children. Altogether a fascinating place not only for the windmill enthusiast but every lover of the countryside.

To reach Outwood, take the A.25 from Reigate and turn right at the village of Bletchingly; the windmill is about four miles along this road.

Reigate Heath (234501)

Cresting the top of a bracken covered mound on Reigate Heath one comes upon what must be England's most unusual Church—a two hundred years old post mill. The roundhouse of the mill, which was last worked in 1868, was converted into a Chapel in 1880 and is effectively furnished. There is an altar with brass candlesticks, an oak reading stand, a small organ and seating for a congregation of fifty people. A choir was dispensed with many years ago, because the small boys who formed it preferred looking for lost balls on the adjacent golf course to hymn practice! Services are held on the third Sunday of each month during the summer—a notice board at the beginning of the path leading to the mill Church provides the details.

Above the roundhouse in the body of the mill, most of the original machinery is still in place. Two pairs of underdrift stones are mounted in the breast and a flour machine is carried at the rear of the stone floor. During its working days the mill was run with patent sails, but these were blown off in 1943. A curious feature is the mock fantail, mounted in the position where the striking wheel should be. This was done during repair work in 1927, probably in error.

Formerly belonging to the Reigate Heath Golf Club and in danger of becoming derelict, the mill was bought by the Borough Council in 1962, and repairs were put in hand which included new sails.

Reigate Heath is situated on the west side of the town just off the A.25.

Shirley (355652)

This well kept tower mill stands in the grounds of the John Ruskin Grammar School on the site of an earlier post mill, which was burnt down in 1854.

The tower, which dates from about 1860, is built of brick and is about fifty feet in height. The post mill shaped cap carries a fantail and the framework of dummy sails. The machinery within the mill is almost entirely of iron, as was usual in the later mills of this type. Two pairs of underdrift stones, one pair peak and the other burr, are still in place plus a wire dresser which is also driven from below via gearing from the great spur wheel.

Inactive since the turn of the century the mill is now an attractive landmark and a link with former days, when Shirley was a small rural community.

Tadworth (236555)

This tall and rather sombre post mill stands in the garden of a private house called, appropriately, Millfield. Windmills have stood here or nearby since the 14th century and a companion to this one was demolished in 1890.

The existing mill has a two storey roundhouse of brick and weatherboarding and was winded by tailpole. The black weatherboarded body has been shorn of sails for very many years—these were of the patent type and drove two pairs of stones, peak and burr, mounted in the breast of the mill. The mill was last worked in 1902, assisted by a stationary steam engine installed in the roundhouse. Although the mill contains very little of the original machinery, its structure is very sound, having been well maintained by its owner.

This is a very easy mill to find, being a prominent landmark from both Banstead Heath and the B.2032 road.

Wray Common (26951)

Said to have last worked by wind in 1895 this picturesquely sited mill is kept preserved as a visual amenity for the locality. Although externally complete with a set of sails and a fantail, no machinery remains within except for the cast-iron windshaft and composite brake wheel. Nowadays the old mill, which was built in 1824, continues to give good service as a store.

Reigate Heath Mill was in danger of becoming just another derelict, but it was saved by the Borough of Reigate Town Planning Committee. New sails have now been fitted and the mill completely renovated.

The windmill on Wray Common. (Photo: A.F. Kersting).

SUSSEX

Argos Hill (571284)

This post mill is of unusual interest in having a tailpole mounted fantail; a rare survival of a one time typical Sussex feature, although curiously enough the only other remaining example is at nearby Cross-in-Hand. In other respects it commands our attention—its setting is its most obvious charm, for the view south from the field in which the mill stands is delightful. The mill is in excellent condition and very distinctive in appearance with its weatherboarding painted white and the sheet-iron roof painted red, an unusual but very effective colour scheme. Four patent sails with shutters removed are fitted and all the machinery, including two pairs of stones and a flour dresser, is intact. The single storey roundhouse is of tarred brickwork. The main post is noteworthy in not being, as is usual, hewn from a single piece of oak, but made up from a number of pieces of pitch pine, a feature which also survives in the post mill at Clayton.

The mill is situated up a lane leading off to the left of the A.267 approximately a mile north of Mayfield.

Downs Mill or Hoads Mill, Bexhill (733088)

This sad ruin is all that remains of a once fine post mill. Stones and gearing and other oddments of machinery lie in disarray where they fell when the mill suddenly collapsed during a violent rainstorm in 1965. The post and trestle still stand quite firmly, almost defiantly, on the roofless roundhouse, so the mill's demise was most probably due to failure of the crowntree. The massive cast-iron windshaft has been placed beside the adjoining bakery and may be inspected

Chailey (North Common) (387214)

This pristine white smock mill is the survivor of two windmills which once stood at Chailey, the other was at South Common, but only its base now remains. The existing mill is kept in good order and carries four sails and a fantail. No machinery remains below the windshaft and the mill these days serves the useful purpose of Boy Scout Headquarters for the lads from the Heritage Craft School, in whose grounds it stands.

Clayton Mills — 'Jack' and 'Jill' (303135)

These two mills are picturesquely situated on a crest of the downs, overlooking the village of Clayton, and are a conspicuous landmark for miles around.

The post mill 'Jill' was originally built in Dyke Road, Brighton in 1821. It was moved to the present site using a team of oxen many years later. It last worked in 1909. The mill body is painted white with a two storey roundhouse tarred black. The mill was originally winded by fantail mounted on the end of the ladder. This is now missing, but remains of the track is still visible. The patent sails originally fitted drove two pairs of stones, one peak and one burr, but the present sails are without shutters. The mill is in very good condition, being maintained by Cuckfield Rural District Council, who were presented with it in 1958 by Mr. Henry Longhurst of golfing fame, who lived in a beautiful modern house nearby.

The tower mill 'Jack' is built of brick with four floors, and is tarred black. It was originally fitted with patent sails which drove the four pairs of stones, two peak and two burr. The cap is painted white and was controlled by a fantail, the remains of which is still in place. This mill was built in 1876 to replace a post mill known as Duncton Mill, built *c.*1770. The roundhouse of Duncton is still standing next to the tower and was used as a study by Mr. Longhurst. E.A. Martin, author of *Life in a Sussex Windmill,* lived here while studying the dewponds.

Two splendid windmills in an idealic situation, easily reached by taking the lane off to the right of the B.2306 going from Pyecombe to Clayton.

Argos Hill post mill. (Photo: A.F. Kersting).

'Jack' windmill at Clayton. (Photo: A.F. Kersting).

'Jill' windmill at Clayton. (Photo: A.F. Kersting).

Cross-in-Hand (558219)

Until 1969 this post mill was one of the few mills still working in England under wind power. However, a broken stock which, in falling, seriously damaged a second sail, has put the mill out of action. The mill has an interesting history, having been moved twice—the first occasion from Uckfield in 1855 to a site nearby, and when moved again to its present position the body was raised onto the existing two storied roundhouse to make more effective use of the wind.

With the exception of the damaged sails the mill is in reasonably good fettle. Two pairs of stones are carried, one pair driven direct from the brake wheel and the other pair underdrift from an iron spur wheel with mortised wooden cogs. The body is covered in white painted sheet-iron, a practice much favoured in Sussex, and the fantail carriage is fitted to the tailpole, another feature at one time common in the county, but only now surviving here and at Argos Hill. This is a very interesting mill and one which it is hoped will be put back into commission in the not too distant future.

Situated near the junction of the A.267 and A.265 roads.

Oldland Mill, Ditchling (321163)

This post mill is one of the oldest in Sussex and appears to be in quite good condition, except for the rear steps and tailpole, which are rather dilapidated. The white painted body carries spring sails which drove two pairs of stones, one peak and one burr, plus a flour machine. The single storey roundhouse is of brick and octagonal in shape. It was once used as a museum by the Sussex Archaeological Society.

Although this attractive mill is visible from the Ditchling–Keymer road it is not easy to find. Leaving Ditchling on the B.2116 take the first turning on the right. After about half a mile the tarmac surface ends at a gate, beyond which is a cart track. The mill is about three hundred yards further along.

Earnley (817983)

A rather ramshackle black tarred smock mill, surprisingly complete however with a full set of sails, a pair of common and a pair of spring. For how long the mill has stood here is a matter for conjecture—local opinion suggests about two hundred and fifty years.

Situated in a private garden beside the B.2198 between Birdham and East Wittering.

Halnaker (920097)

A preserved shell carrying a new cap and sails. This is the oldest tower mill in the county, being built in 1740. It does not however contain any machinery whatever and there is therefore not much of interest to warrant a special visit, except for the fact that, perched on top of the downs, this mill commands a delightful view of the surrounding countryside.

Situated at the top of a downland track leading off the A.285.

High Salvington (123067)

This well preserved post mill provides a reminder that this residential district was once open downland. It has a black body, the framework of four common sails and a tailpole.

Exactly when the mill was built is unrecorded but was probably during the early 18th century. It last worked in 1897. Machinery inside the mill includes two pairs of overdrift stones, mounted head and tail fashion. A massive wooden windshaft carries a tail-wheel, also wood, but the brake wheel is missing. The rest of the gear, quants, stonenuts, governors, bolter reel etc. are present in the mill but not, unfortunately, fitted up.

Windmill Hill, Hurstmonceaux (648122)

Large derelict post mill. Body covered with very rusty iron sheeting. Sails and most of the machinery removed. The date of erection is not known, but it is possibly the mill indicated on the first large scale map of Sussex made by Richard Budgeon in 1724. It ceased work shortly before the First World War. A curious feature of the mill is the roundhouse, a two floored affair having the lower part about twice the diameter of the upper, giving the mill a unique appearance.

Prominently sited beside the A.271 about a mile east of Hurstmonceaux.

Hogg Hill Mill, Icklesham (887160)

This fine post mill with tarred body and white sails is a distinctive landmark in the lovely countryside at Icklesham.

The two storied brick roundhouse is tarred black. The sails are of the spring type and worked two pairs of stones and a flour machine. The most unusual feature of the mill is the position of the fantail, which is mounted on the rear of the roof. This position necessitates a long drive shaft and gearing to the wheels at the foot of the ladder. A crank handle is provided at ground level to wind the mill by hand.

The mill is reputed to be nearly three hundred years old and was moved to its present site from nearby Pett Hill approximately one hundred and fifty years ago. This magnificent post mill in its beautiful setting must be visited by the keen Windmillian.

With Hastings behind you on the A.259, take the road on the right at Icklesham leading to Pett, the mill is on the left-hand side of the road and cannot be missed.

Nutley (451291)

This windmill stands in a clearing in the beautiful Ashdown Forest, and is said to be nearly three hundred years old. It is a small post mill with an open trestle, one of a handful which have survived in England. It has recently undergone a very extensive restoration scheme. The trestle has been completely renewed, a broken side girt replaced, and the crowntree, which was found to be badly rotted, has been hollowed out and strengthened by the insertion of a steel joist. The final task was the fitting of four common sails. The mill ceased working in 1908 and was then already in poor condition. Only the timely repair work put in hand in 1929 by the then new owner, Lady Castle Stewart, saved it from collapse.

All the machinery is still in place, two pairs of stones are mounted head and tail fashion, driven direct from the brake and tail-wheels which are made of wood, the stonenuts being iron. The very worn rim of the tail-wheel suggests that it may formerly have been a brake wheel either in this mill or another. The original wooden windshaft has at sometime been replaced by the present cast-iron shaft, which was evidently from another and larger mill, the tail-wheel being mounted some distance in front of the squared section on which it would normally be carried. The external appearance of the mill is primitive but attractive, the mansard shaped roof is uncommon and is possibly an intermediate development between the straight pitch of the earlier mills, such as Bourn, and the later curved shapes still to be seen on most of the surviving post mills. When the wind is favourable the grinding performance of the mill may be demonstrated.

The mill is situated in delightful countryside near Nutley village, and stands at the end of a narrow track which leads off to the left of the minor road from Nutley to Crowborough.

Nyetimber (892989)

A derelict tower mill without sweeps or fantail which is almost completely hidden under festoons of ivy. Said to have been built in the early 19th century, it last worked under wind power in 1915.

Polegate (582041)

This forty-seven foot high tower mill, which was last worked by wind in 1943, is in excellent condition, having been restored to full working order by the Burgess Hill millwrights Edwin Hole and Son, on the instructions of the Eastbourne and District Preservation Society. At one time threatened with demolition, the mill was saved by the exceptional generosity of Mr. A. Ovenden, the last miller, who in 1962 offered the mill and site to the society for £1,000, which was less than half its market value. An appeal for £3,500 was launched by the society to purchase and renovate the mill. The response to this appeal was such that by 1965 sufficient funds were available to enable the repair work to begin and renovation was completed by 1968.

Built in 1817, the body which is of red brick and partly tiled has a dome shaped cap carrying four patent sails. It is winded by a five bladed fantail. There are three pairs of stones, two pairs being overdrift and the third pair, a later addition to the mill, are underdrift. On the spout floor is a flour machine and a smutter, both belt driven. The doors on this floor provided access to the former outside stage, no vestige of which remains.

In the adjoining outbuilding is an interesting collection of bygones and photographs, relevant to milling history. Particularly fascinating is the beautifully constructed model of a typical Sussex post mill made by Mr. F.W. Gregory.

Situated just off the A.22 at Polegate, this very interesting mill is open to visitors during the summer months, when the sails may be seen turning if both wind and volunteers to man the mill are available in sufficient strength.

Punnetts Town (627209)

This lovely white smock mill is in working order and is an outstanding example of how successful a restoration project can turn out, providing the skills and enthusiasm are available. Looking over the mill today it seems almost unbelievable that Mr. Dallaway, whose family have had a milling business here for generations, started work on what was, in 1947, virtually an empty shell. A new cap and brake wheel had to be specially made and machinery was acquired from other mills as they were dismantled. A full set of sails and a fantail completed the restoration.

The mill has an interesting history. It was erected here in 1856 to replace a post mill which was destroyed by fire. Prior to that date it stood at Biddenden in Kent, where it was known by the curious name of Cherry Clack windmill.

The mill is situated on the edge of the village in delightful surroundings.

Rottingdean (366025)

This rather forbidding looking smock mill is a well-known landmark standing as it does so prominently on top of the cliffs. The exact date the mill was built is not known but, in common with most mills situated close to the sea, local tradition links the mill's early days with smuggling.

There is no machinery left inside, but the mill is maintained by the local authority as a visual amenity. Unfortunately it suffers from the attentions of local vandals who find its black tarred exterior an eminently suitable surface for chalked slogans.

Being in such an exposed situation, and subjected to many decades of strong winds, the mill had developed a curious twisted lean which gave it an unsightly appearance. This defect has recently been corrected by millwright Edwin Hole by the provision of a supporting framework of reinforced steel joists inside the structure, which should resist all future attempts by the wind to blow it over.

Selsey Medmerry Mill (844935)

A brick tower mill with four sails, but of little interest, having been stripped of all its machinery and converted into a shop serving a nearby caravan site.

Stands, preserved as a landmark, in rather drab surroundings near Selsey Bill.

Facing page: The forbidding looking smock mill above the cliffs at Rottingdean. (Photo: A.F. Kersting).

Kings Mill, Shipley (140220)

With a fine square towered 12th century Church and the river Adur winding its way through lush green meadows, the setting of this windmill must surely rank as one of the loveliest in England. The mill has been restored to working order as a memorial to Hillaire Belloc the writer, who lived here from 1906 until his death in 1953. Restoration was carried out in 1958 by Edwin Hole of Burgess Hill and the mill has been subsequently kept in splendid condition.

Built in 1879, it is the youngest and largest windmill in Sussex. It is outstandingly handsome, a white painted octagonal smock with a two storied tarred brick base. Four double shuttered patent sails drive three pairs of underdrift stones and a wire machine. The cap, which is conical in shape, has a prominent fantail and below the fan staging is carried the framework for the striking wheel. The striking chain hangs down to the outside gallery at second floor level. From this gallery the miller would adjust the shutters of the sails by hanging suitable weights on the chain. The machinery inside the mill is in good order and is mostly made of iron, except the brake wheel and the upright shaft—these are made of timber and were probably brought here from an earlier mill. A short portable ladder on the dust floor provides a means to inspect the machinery in the cap at close quarters and also enables the more intrepid to venture out onto the fan staging. Access to the fan stage is via a small outward opening door in the tail of the cap. From this lofty and windswept perch some sixty feet above the ground a magnificent panorama of the surrounding countryside is revealed, which in itself merits a visit. In a favourable wind the sweeps and machinery may be seen running.

Shipley is just off the A.272 about four miles east of Billingshurst.

Facing page: The meadows and windmill at Shipley. (Photo: A.F. Kersting).

Stonecross (620044)

A white painted cement rendered tower mill in good state of repair. This mill was working under wind power up until just before the war with a single pair of patent sails—these are still intact along with the fantail. The cap is beehive shaped, a type once common in Sussex. All the machinery is in place, including three pairs of overdrift stones, a flour machine and a smutter. A distinctive feature of this mill are the circular wheel-like windows in the tower, repeated in the large diameter single storied base as half circles, very unusual.

Situated beside the A.27 midway between Polegate and Westham.

Stonecross Mill. This picture was taken before the mill lost two of its sails in a gale in 1925.

West Blatchington (277067)

An interesting small smock mill perched on top of a flint barn. The body is hexagonal in plan and covered with weatherboarding. Four patent sails and a fantail are carried and most of the machinery is intact. Exact date of building is not known, but a mill symbol is shown about here on Budgeon's large scale map of the county dated 1724.

Winchelsea (901177)

Large post mill body on single storey brick roundhouse. All machinery removed and nowadays used for storing hay. Built around 1760, last worked about 1908. The fantail, long since gone, was mounted on the roof in a similar fashion to that at Icklesham post mill, which can be seen across the valley from here.

Above: The smock mill at West Blatchington in 1934. This mill and the adjoining barn were used by smugglers to hide their goods. The barn was gutted by fire in 1936, as shown in the photograph below. The windmill has since been restored and carries four new sails and a fantail.

Kings Mill, Shipley, Sussex — Built in 1879, this mill is preserved in working order as a memorial to Hilaire Belloc the writer, who lived in the Mill House from 1906 until his death in 1953.

Syleham Mill, Suffolk — Built in 1730, this post mill was brought here from its original site around 1823.

Mountnessing Mill, Mountnessing, Essex — Built in 1807, this fine old post mill is being preserved by the County Council.

Hogg Hill Mill, Icklesham, Sussex — Built during the latter part of the 17th century, this post mill is unusual in having a roof mounted fantail, a rare survival.

Friston Mill, Suffolk — Built in the early part of the 19th century, this is the tallest surviving post mill in the country.

Outwood Mill, Surrey — Built in 1665, this post mill is England's oldest working windmill and may be seen working at weekends in the summer if the wind is favourable.

Chailey Mill, Sussex — Built *c.*1830, this attractive smock mill was erected on this site in 1864, having previously stood at Newhaven and earlier still at Highbrook.

Nutley Mill, Ashdown Forest, Sussex — This photograph shows this post mill shortly before it was restored to working order. The mill is of special interest, in having an open trestle, the only example left in Sussex.

WARWICKSHIRE

Chesterton (348594)

This unusual stone tower mill stands prominently on a hill a short distance from the Roman Camp at Chesterton. It is reputed, on slender evidence, to have been built originally as an observatory in 1632 for Sir Edward Peto. Fully renovated in recent years—it was completed in 1971—the exterior is unique in appearance, the two floors of the mill being supported on six semi-circular arches, access being via a trap door in the lower floor. Little machinery remains, but of uncommon interest are the compass arm brake wheel and a primitive wallower which is probably the only surviving lantern pinion in an English windmill.

Chesterton is situated on the Fosse Way, the Roman road stretching from Devon to Lincoln and the North.

Norton Lindsey (224633)

Stocky brick tower mill with single gabled cap and two common sails.

WILTSHIRE

Wilton (276617)

This windmill at Wilton is the only one left standing in the county, although a hundred or so years ago they were fairly numerous. The importance of its link with Wiltshire agricultural history led to its aquisition in 1971 by the County Council, and it has since been completely restored to working order. It is a five storied brick tower, built in 1821 to replace the watermill at Great Bedwyn which was destroyed in 1800. The mill worked until the 1920's when the estate on which it stands was sold. Despite its long years of disuse, the structure was found to be fairly sound when renovation was put in hand and all its machinery, including three pairs of underdrift stones, substantially intact. New sails, cap and fantail have been fitted and the mill can be seen working during summer weekends when the wind is favourable.

Situated prominently on a chalk ridge, a visit to this interesting mill can be combined with a visit to the nearby Crofton pumping station with its steam driven beam engines on the Kennet and Avon canal, built in 1810.

WORCESTERSHIRE

Danzey Green, Avoncroft Museum of Buildings, Near Bromsgrove

Originally built during the late 18th century, this splendidly reconstructed post mill had stood at Danzey Green, near Tamworth-in-Arden, until 1969 when, despite (or perhaps because of) its rickety state, it was carefully dismantled and brought to Avoncroft.

It is a type of post mill once common in the Midlands and the peculiarity of its design is the track or curb on top of the roundhouse wall around which rollers ran to help support the weight of the mill body. Although at present without machinery the rest of the mill, including sails and wooden windshaft, has been completely re-built; the roundhouse by boys from Hewell Grange Borstal using the original bricks. The mill can be brought into wind by means of the tailpole. The sails, one pair common, one pair spring, will idle in a moderate breeze.

Avoncroft is situated at the junction of the A.4024 and B.4091 roads near Bromsgrove and the mill is just one of several interesting buildings which have been saved from destruction and re-built on the museum site. Others include a pair of iron age dwellings, a 15th century timber framed house, a chainmakers workshop, a nailmakers workshop and a granary, altogether a place not to be missed.

Windmills Lost and Decayed

Every so often a treasure-trove of old photographs is found that convey so much more readily than mere words the quality of life in years gone by. Sadly, many photographs of old mills have been lost for ever and a great many of our grandfathers' photographs would have shown some of those windmills that are now gone. There are probably a great many photographs and records gathering dust—awaiting some budding historian or archivist who will appreciate their value.

Here are just a few which have, alas, survived longer than the mills they depict.

The windmill at Hawkinge, Nr. Folkestone, Kent, 1935. Nothing remains today of this old smock mill.
A new fantail was fitted in 1890, but by 1902 the mill ceased working and, like so many Kentish windmills, was allowed to fall into disrepair.

Below: Town Mill, Biddenden, Kent. Believed to be one of the oldest smock mills in Kent. Note the unusual shape of the cap, which overhung the body more than usual. When this photograph was taken in 1937 the mill was already beginning to deteriorate, but could have been saved. Nothing was done and the mill continued to decline as the photograph on the facing page shows. By 1940 it had lost its fantail and the structure was in a dangerous state.

The Old Mill, Grange Road, Ramsgate, Kent. The sequence of pictures on this page illustrate the decline and final fall of the once fine black smock mill. By 1930 the base alone remained and this was covered in advertisement posters and it was used as a garage.

The post mill, Smarden, 1926, believed to have been the oldest mill in Kent.

The smock mill, Smarden, 1926. The mill was last worked in 1923.

Left: Newington Mill, Nr. Ramsgate, *c.* 1930.

Delce Mill, the last of many mills in the City of Rochester, Kent.

St.Martin's Mill, Canterbury, a stone tower mill which ceased working in 1890 and was converted to a residence in 1920.

Right: Details of the sweeps of the mill at Betteshanger, Nr. Deal, Kent.

Below right: The body of the Upper Mill, Woodchurch, Kent. This was demolished in the 1930's.

Below: The tower mill at Stanford, Kent, 1930, and in working order.

Above: The brick tower of Edenbridge Mill, Kent, 1933. Note the unusual 'bee-hive' cap.

Right: The old post mill at Frittenden, Kent, 1920. By 1930 the sweeps were gone and the mill was derelict.

Below right: The tarred smock and stage mill at Milton, Nr. Sittingbourne, Kent. The mill became derelict and was burnt down soon after the turn of the century.

Below: Bridge Mill, Bridge, Nr. Canterbury, Kent, 1927.

The mill at Bull Lane, Boughton, Nr. Faversham, 1920, once the second largest mill in Kent.

By 1930 only the weatherboarded tower of Boughton mill remained.

The mill at Preston, Nr. Faversham, Kent. One sweep broke off in 1914 and the remaining three were removed soon after.

The smock mill at Sutton Valence, the second tallest in Kent, which ceased working in 1918. By 1930 the sweeps had been removed and only the frames remained.

The fine smock mill at Upper Deal, Kent, 1926. This mill was demolished in 1929, when the land was required for homes for colliery workers.

Boundgate Mill, Badlesmere, Nr. Faversham, Kent, 1926.

Charing Old Mill, Kent. The sweeps were removed in 1917 after being damaged in a high wind.

Jarvis Mill, Doddington, Kent. Ceased working in 1926 and then deteriorated rapidly.

Peasmarsh windmill, Sussex. Old and decrepit by 1935—a sorry sight!

Mowbray's Mill, Pocklington, Lincs., 1935.

Downfield Mill, Soham, Cambs. Re-built in 1890 as a tower mill, although originally a smock, this mill was working until the mid-1950's, but it is now derelict.

Little Mill and Great Mill, Prospect Hill, Frindsbury, Kent. The Little Mill was struck by lightning in 1886 and then demolished. Great Mill was pulled down in 1890 and the base converted into a pair of cottages.

Conningsby Mill, Lincs., 1935. One of the few mills to have five sweeps.

The octagonal smock mill at Great Chart was one of only two which had six sweeps. It was pulled down in 1928 and the base used as a store.

Lower or Great Mill, Woodchurch, Kent, last worked in 1926.

Bredcar Mill, Kent. By 1909 the mill cap and body were being used for firewood!

Stocks Green Mill, Leigh, Nr. Hildenborough, Kent, one of England's oldest smock mills; last worked in 1910.

The smock mill at New Romney, Kent, 1905.

Above: Silver Hill Mill, Hastings, Sussex. This splendid white smock mill was working in 1928 when this picture was taken.

Opposite: Higham Mill, Nr. Rochester, Kent, pulled down after being damaged in a gale in 1920.

Below: The remains of the unusual mill on Littleworth Heath, Oxon.

Below: The remains of Nonington smock mill, Nr. Dover. The body was covered in heavy-gauge iron sheeting.

White Mill, Lyminge, Kent. One of two smock mills in this area. The second mill, known as Black Mill, was destroyed by fire in 1891. White Mill was demolished in 1920 after it had fallen into a state of disrepair.

Reed Mill, Kingston, Nr. Canterbury, Kent. The mill was badly damaged in 1915, when the cap and sails were swept away in a snowstorm.

Rosby Mill, Lincs., 1935.

Appendix — Windmills Open to Visitors

Listed below are windmills which are open to the public by courtesy of the various owners. In some cases a small charge is made to view the interior, in others a collection box may be present inside the mill. In both instances the income derived helps with the costly business of maintaining the mill. Most windmills stand on private land and all windmills *belong to someone,* so please co-operate with the person(s) in charge of the mill with regard to car parking etc. Particular care should be taken to avoid trampling on growing crops.

The list of mills which follows is gratifyingly long and reflects the upsurge of interest that has taken place in recent years. Consideration for both the privacy and property of people living near windmills will ensure that access to these interesting old structures continues.

N.B. Viewing arrangements, where stated, apply at the time of writing, but visitors are advised to confirm that these still apply before undertaking a long trip to visit a mill.

BEDFORDSHIRE
Stevington Post Mill — Open by arrangement with Bedfordshire County Council. Apply to Clerk of Council, County Hall, Bedford.

BUCKINGHAMSHIRE
Brill Post Mill — Open Sunday afternoons April to September.
Pitstone Green Post Mill — Open Sunday afternoons during the summer months.

CAMBRIDGESHIRE
Bourn Post Mill — Open every day during the summer.
Great Chishill Post Mill — Key to the mill may be obtained locally.
Over Tower Mill — One of the few mills which can be seen running. The interior may be viewed at Bank Holidays and at other times by arrangement with the owner, Mr. G.C. Wilson, Mill Villa, Longstanton Road, Over.

ESSEX
Aythorpe Roding Post Mill — Viewing by arrangement with Essex County Council, County Hall, Chelmsford.
Bocking Post Mill — Open all the year round. Key available at Public House opposite.
Mountnessing Post Mill — Viewing of the interior by arrangement with Essex County Council, County Hall, Chelmsford.
Stanstead Mountfitchet Tower Mill — Open to view from 2.30 to 7.00 p.m. on the first Sunday in each month from April to October, and on Easter, Spring and Summer Bank Holiday weekends.
Stock Tower Mill — Interior may be viewed by arrangement with Essex County Council, County Hall, Chelmsford.
Upminster Smock Mill — Open at specified times.

KENT
Chillenden Post Mill
Union Mill (Smock), Cranbrook
Drapers Smock Mill, Margate — Open Sunday afternoons during summer months.
Meopham Smock Mill — Open Sunday afternoons July and August.
West Kingsdown Smock Mill

LEICESTERSHIRE
Kibworth Harcourt

LINCOLNSHIRE
Alford Tower Mill — Open on specified days during the summer months.
Burgh-le-Marsh Tower Mill — Open to inspection at your own risk. Key available from the adjacent windmill restaurant.
Heckington Tower Mill — Visitors may explore the interior at their own risk.
Wrawby Post Mill — Open Spring and Summer Bank Holiday weekends.

LONDON
Brixton Tower Mill — Open Saturdays and Sundays May to September.

NORFOLK
Berney Arms Drainage Mill, Reedham — Open April to September 9.30 a.m. to 7.00 p.m. **N.B.** Can only be reached by boat or rail.
Billingford Tower Mill — Apply to Norfolk County Council, Planning Department.
Horsey Mere Drainage Mill — Left open during the summer months. Visitors may explore the interior at their own risk.
Stracey Arms Drainage Mill — Left open during the summer months. Similar to Horsey Mill, visitors may explore at their own risk.

NOTTINGHAMSHIRE
North Leverton Tower Mill — Open on specified days in summer months.

SOMERSET
Ashton Tower Mill, Chapel Allerton — Apply to Bristol City Museum.
Stembridge Tower Mill, High Ham, Langport — Open to view most Sundays.

SUFFOLK
Drinkstone Post Mill
Framsden Post Mill — Open Spring and Summer Bank Holiday weekends.
Herringfleet Smock Mill — Open by arrangement. Write to Clerk of the East Suffolk County Council, County Hall, Ipswich.
Holton Post Mill, Holton St. Peter, Nr. Halesworth — Open Spring and Summer Bank Holiday weekends.
Pakenham Tower Mill — Open all day every weekday 9.00 a.m. to 6.00 p.m.
Saxtead Green Post Mill — Open May to September every day except Sunday.

SURREY
Outwood Post Mill — Open on Sunday afternoons.

SUSSEX
Argos Hill Post Mill — May be inspected by arrangement with the Engineer and

Surveyor, Uckfield Rural District Council.
High Salvington Post Mill — Open every day.
Nutley Post Mill — Open on specified Sundays during the summer months.
Polegate Tower Mill — Open Sundays and Bank Holidays, May to September.
Shipley Smock Mill — Open first weekend of each summer month, including Bank Holidays.

WORCESTERSHIRE
Bromsgrove, Avon Croft Museum of Ancient Buildings (The post mill from Danzey Green) — Open on Saturdays and Sundays April to September.

The expertly restored post mill at Saxtead Green, Suffolk—a mill open to the public.

Bibliography

General Books

Batten, M.I. *English Windmills, Vol. 1.* Containing a history of their origin and development, with records of mills in Kent, Surrey and Sussex. (Architectural Press, 1930)

Brangwyn, Frank and Preston, Hayter. *Windmills* (John Lane, 1923)

De Little, R.J. *The Windmill, Yesterday and Today.* (John Baker, 1972)

Farries, K.G. and Mason, M.T. *The Windmills of Surrey and Inner London.* (Charles Skilton, 1966)

Finch, William Coles. *Watermills and Windmills.* A Historical Survey of their Rise, Decline and Fall, as portrayed by those of Kent. (C.W. Daniel, 1933)

Freese, Stanley. *Windmills and Millwrighting.* (Cambridge University Press, 1957). (Reprinted David and Charles, 1971)

Hemming, Peter. *Windmills in Sussex.* A description of the construction and operation of windmills, exemplified with up-to-date notes on the still existing windmills in Sussex with photographic illustrations. (C.W. Daniel, 1936)

Hopkins, R. Thurston. *Old English Mills and Inns.* (Cecil Palmer, 1927)

Hopkins, R. Thurston. *Old Watermills and Windmills.* (Philip Allen, n.d.)

Hopkins, R. Thurston and Freese, Stanley. *In Search of English Windmills.* (Cecil Palmer, 1931)

Mais, S.P.B. *England of the Windmills.* (J.M. Dent, 1931)

Reynolds, John. *Windmills and Watermills.* (Hugh Evelyn, 1970)

Smith, Donald. *English Windmills, Vol. 2.* Containing a record of the mills in Bucks., Essex, Herts., Middx. and London. (Architectural Press, 1932)

Wailes, Rex. *Windmills in England.* A study of their origin, development and future. (Architectural Press, 1948)

Wailes, Rex. *The English Windmill.* (Routledge and Keegan Paul, 1954)

Booklets

Adams, W.R. *Windmills in Kent.* (Kent County Council, 1955)

Butler, Anthony D. *Upminster Mill.* (Peter Davis, 1968)

Fowell, G.M. *Windmills in Sussex.* (Walker Galleries, 1930)

Hopkins, R. Thurston. *Windmills.* (Charles Clark, 1931)

Laver, James and Wailes, Rex. *Windmills.* (Society for the Protection of Ancient Buildings)

Paddon, J.B. *Windmills in Surrey.* (Oxonian Press, 1926)

Paddon, J.B. *Windmills in West Sussex.* (Oxonian Press, 1926)

Paddon, J.B. *Windmills in East Sussex.* (Oxonian Press, 1926)

Paddon, J.B. *Windmills in the Midlands.* (Oxonian Press, 1926)

Paddon, J.B. *Windmills in Kent.* (South Eastern Gazette Press, 1926)

Vince, J.N.T. *Discovering Windmills.* (Shire Publications, 1969)

Newcomen Society Transactions.

Glossary

Annular Sail — Wheel-like sail with a single row of radial shutters.
Bed Stone — The lower of a pair of mill stones.
Bell Alarm — Device to indicate hopper needs refilling with grain.
Bill — Stone dressing chisel.
Bin Floor — Upper floor where the grain storage bins are situated.
Bolter — Machine for separating flour from the bran and middlings.
Brake — Band which contracts onto rim of brake wheel.
Brake Wheel — Large cogged wheel on fore-end of windshaft.
Brayer — Lever which actuates the bridge-tree.
Breast — Fore-end of post mill body.
Breast Beam — Carries the neck bearing for the windshaft.
Bridge-tree — Pivoting beam carrying the stone spindle bearing.
Bridging Box — Bearing for lower end of stone spindle with provision for sideways adjustment.
Buck — Suffolk term for post mill body.
Burr Stone — Stone imported from France.
Cannister (Poll End) — Iron casting at fore-end of windshaft, through which the sail stocks are fixed.
Cant Posts — Inclined structural members at the corners of a smock mill.
Cap — Rotating cupola on top of a smock or tower mill.
Cap Circle — Wooden frame under the cap carrying the rollers or skid plates.
Clasp Arm Wheel — Wooden gear wheel with spokes that form a square opening into which the shaft fits.
Cloth Sails — Sails spread with canvas. Syn. common sails.
Cock Head — Top part of a stone spindle.
Collar — The lower bearing of a post mill which steadies the body.
Common Sails — Sails spread with canvas. Syn. cloth sails.
Compass Arm Wheel — Wooden gear wheel with spokes mortised into a wooden shaft.
Composite Mill — Post mill body adapted to a squat tower.
Cracks — The fine grooves cut in a mill stone.
Crook String — String for adjusting the inclination of the grain shoe feeding the stones.
Cross — Heavy cruciform casting on fore-end of windshaft to which the sails are clamped. (Alternative to poll end).
Crosstree — Horizontal tie beams of a post mill trestle.
Crotch — Forked lower end of a quant.
Crowntree — Main horizontal beam which pivots on top of the post in a post mill.
Cullin Stone — Stone imported from Cologne.
Curb — Track on smock or tower mill on which cap rotates.
Damsel — Multi-armed spindle which projects above underdrift stones and vibrates the grain shoe.
Dead Curb — Curb with greased skid plates instead of rollers.
Double Shuttered — Sail with shutters on both sides of the whip.
Dust Floor — Top floor of smock or tower mill.
Eye — Hole in the centre of runner stone into which grain is fed.
Fan Stage — Staging below the fantail.
Fantail — Multi-vaned wheel to turn cap or body into wind automatically.
Footstep Bearing — Thrust bearing for an upright shaft.
Furrows — Large grooves cut in a mill stone.

Gallery — Staging around the cap of a smock or tower mill.
Gate — Sliding plate which regulates the flow of grain from the shoe.
Glutbox — Upper bearing box for a quant which allows sideways movement to disengage the drive.
Governor — Device to automatically maintain the gap between the stones.
Great Spur Wheel — Cogged wheel on upright shaft which drives the stone nuts.
Grist — Corn to be ground.
Gudgeon — Iron journal fixed to wooden shaft.
Gymbal Bar — Iron bar set across the eye of the runner stone and which locates in the mace atop the stone spindle.
Head Sick — Post mill with the body leaning forward.
Hollow Post Mill — Post mill which has stones located in lower part of mill, driven by a shaft passing down through the centre of the hollow post.
Hopper — Grain container which feeds the stones.
Horse — Frame on which the hopper is mounted.
Hurst — Raised framework carrying the stones.
Jog-Scry — An oscillating sieve.
Keep Flange — Horizontal flange under which truck wheels run to keep cap central on the curb.
Lands — The raised areas between the furrows on a stone.
Lantern Pinion — Gear wheel made up of two flanges spaced apart and connected by dowels around the circumference which act as cogs.
Leading Board — Fitted to leading edge of sail.
Live Curb — Curb on which cap rotates on rollers.
Mace — Small casting which fits on top of stone spindle and carries gymbal bar.
Meal Floor — The floor immediately below the stone floor on which the meal spouts are located.
Meal Spout — Canvas chute from which the meal from the stones is fed into sacks.
Mill Bill — Chisel used for stone dressing.
Neck Bearing — Bearing for the front end of the windshaft.
Overdrift — Stones driven from above.
Patent Sail — Sail with shutters which can be adjusted whilst in motion.
Peak Stone — Stone quarried in Derbyshire.
Petticoat — Overhanging weatherboarding on cap or post mill body to keep out rain.
Piers — Brick columns supporting the crosstrees of a post mill.
Pintle — Journal projecting from the top of a post or the tail-end of a windshaft.
Pit Wheel — Cogged wheel mounted on a scoop wheel shaft in a drainage mill.
Poll End (Cannister) — Box-like casting on the windshaft in which the stocks of the sails are fixed.
Post — The main structural timber in a post mill.
Prick Post — Central vertical timber in the breast of a post mill body.
Quant — The final driving spindle to an overdrift stone.
Quarterbars — Diagonal struts mortised into the post and crosstrees of the trestle of a post mill.
Quartering — Turning the mill at right angles to the wind.
Rack — The row of cogs on (a) a curb, (b) on the rear of a striker rod.
Roundhouse — Lower compartment of a post mill enclosing the trestle.
Runner Stone — Upper rotating stone of a pair of mill stones.
Rynd (Gymbal Bar) — The metal bar cemented across the eye of the runner stone.
Sack Hoist — Bollard for lifting and lowering sacks.
Sail In — Close the shutters.
Sail Out - Open the shutters.

Samson Head — Castings mounted on top of the post and the underside of the crown-tree.
Scoop Wheel — Wheel on drainage mill which lifts the water and discharges it at a higher level.
Sheers — (a) Two parallel beams supporting weather and tail-beams in a cap. (b) Two parallel beams beneath a post mill body.
Shoe — Grain trough from the hopper which feeds into the eye of the runner stone.
Shutters — Hinged flaps in sails.
Shutter Bars — A rod running the length of a sail to which all the shutters are connected.
Side Girts — Horizontal structural timbers fixed to each end of the crowntree in a post mill.
Single Shuttered — Sails with shutters on trailing side of whip only.
Smock Mill — Tower mill built of wood.
Soke — Feudal custom whereby grain had to be ground in the manorial mill.
Sprattle Beam — Horizontal beam on which bearing for a vertical shaft is mounted.
Spider — Cruciform coupling which actuates shutter bars in patent sails.
Spring Sail — Sail in which shutters are held shut by spring tension on the shutter bars.
Stage — Platform, usually at meal floor level, around body of a smock or tower mill.
Steelyard — An iron lever connecting the centrifugal governor to the brayer.
Stock — Stout timber through poll end to which sails are fixed.
Stone Floor — On which the stones are located.
Stone Nut — Pinion driving the runner stone.
Stone Spindle — On which the runner stone is mounted.
Storm Hatch — Small door above fore-end of windshaft allowing access to poll end.
Striking Gear — Mechanism for operating shutters in a patent sail.
Striking Rod — Carried in hollow windshaft and is part of the striking gear mechanism.
Sunk Post Mill — Early type of post mill in which sub-structure was buried in the ground.
Sweep — Kentish term for sail.
Tail — The rear end of a post mill body.
Tail-beam — On which the rear bearing for the windshaft is mounted.
Tailpole — Beam projecting from rear of mill to enable the body or cap to be turned to face the wind.
Tail-wheel — Cogged wheel carried on rear end of windshaft in a post mill.
Talthur — Lever to raise post mill ladder off the ground when winding the mill.
Tentering — Setting the gap between the stones.
Toll — Proportion of grain appropriated by the miller in payment for grinding.
Tower Mill — Mill built of brick or similar material.
Trestle — Sub-structure of a post mill.
Trundle Wheel — Early gear wheel with dowels instead of cogs.
Underdrift — Stone driven from below.
Upright Shaft — Driving shaft carrying the wallower and great spur wheel.
Vat — Wooden casing containing the mill stones.
Wallower — Gear wheel on upright shaft meshing with brake wheel.
Weather — The twist in a sail.
Weather-beam (Breast Beam) — Timber which carries the neck bearing for the wind-shaft.
Whip — Timber fixed to sail stock on which the sail framework is built up.
Winding — The operation of turning the mill to face the wind.
Windshaft — Horizontal shaft carrying the sails.
Wire Machine — Machine to extract flour from the meal in varying grades of fineness.

Index